Yamaha FS1E, FS1 & FS1M Owners Workshop Manual

by Mervyn Bleach
with an additional Chapter on the 1975 on models
by Jeremy Churchill

Models covered
FS1E. 49cc. November 1972 to July 1977
FS1E-DX. 49cc. May 1975 to July 1977
FS1E-A. 49cc. April to July 1977
FS1E-DXA. 49cc. April to July 1977
FS1/FS1M. 49cc. August 1977 to October 1979, March 1987 on
FS1-DX/FS1M-DX. 49cc. August 1977 to April 1983
FS1SE. 49cc. April 1981 to October 1983

ISBN 978 1 85010 677 7

J H Haynes & Co. Ltd.
Haynes North America, Inc

www.haynes.com

British Library Cataloguing in Publication Data
Bleach, Mervyn
Yamaha FS1E, FS1 & FS1M owners workshop manual.
– 4th ed.
1. Motorcycles. Maintenance & repair
I. Title II. Churchill, Jeremy, *1954–* III. Series
629.28775
ISBN 1-85010-677-0

T0327551

Acknowledgements

Our grateful thanks are due to Lawrie Hockley of W and H Brockliss Limited, 334 Brockley Road, London, S.E.4 who provided much useful information based on their wide experience as Yamaha repair specialists, and to Jim Patch of Yeovil Motorcycle Services Ltd, Yeovil, Somerset, who provided additional information. Also to Mitsui Machinery Sales (UK) Ltd who supplied the necessary service information and gave permission to reproduce many of the line drawings used. The machine shown on the front cover was supplied by Atkins Motors of Taunton, Somerset.

We would also like to thank the Avon Rubber Company, who kindly supplied information and technical assistance on tyre fitting, NGK Spark Plugs (UK) Ltd, for information on spark plug maintenance and electrode conditions, and Renold Ltd, for advice on chain care and renewal.

About this manual

The author of this manual has the conviction that the only way in which a meaningful and easy to follow text can be written is first to do the work himself, under conditions similar to those found in the average household. As a result the hands seen in the photographs are those of the author. Even the machines are not new; examples that had covered a considerable mileage were selected so that the conditions encountered would be typical of these found by the average owner. Unless specially mentioned, and therefore considered essential, Yamaha special service tools have not been used. There is invariably some alternative means of loosening or removing a vital component when service tools are not available but risk of damage should always be avoided.

Each chapter is divided into numbered sections. Within these sections are numbered paragraphs. Cross reference throughout the manual is quite straightforward and logical. When reference is made "See Section 6.10" it means Section 6 paragraph 10 in the same chapter. If another chapter were meant, the reference would read "See Chapter 2 Section 6.10". All the photographs are captioned with a section/paragraph number to which they refer, and are relevant to the chapter text adjacent.

Figures (usually line illustrations) appear in a logical but numerical order, within a given chapter. Fig. 1.1 therefore, refers to the first figure in chapter one.

Left-hand and right-hand descriptions of the machines and their components refer to the left and right of a given machine when the rider is seated normally.

Motorcycle manufacturers continually make changes to specifications and recommendations, and these, when notified, are incorporated into our manuals at the earliest opportunity. Whilst every care is taken to ensure that the information in this manual is correct no liability can be accepted by the authors or publishers for loss, damage or injury, caused by any errors in or omissions from the information given.

Introduction to the Yamaha FSIE

The Yamaha FS1E was introduced to the UK in November 1972 to comply with the new legislation restricting sixteen year old riders to mopeds. Rather than waste time and money in developing a new purpose-built model, Yamaha adapted an existing design, fitting it with pedals so that it became a moped for all practical purposes.

As the FS1E was based on a proper motorcycle, its performance and handling, and arguably more important, its appearance were far superior to anything available at the time except for a few expensive and temperamental European models. It became an instant success, proving to be the most popular choice with the majority of sixteen year olds and as such is the original 'sports moped'. Its ease of maintenance and reliability have ensured that it is still a favourite in spite of increasing competition.

Contents

Left-hand side view of the 49 cc Yamaha FS1E (394)

Right-hand side view of the 49 cc Yamaha FS1E (394)

Ordering spare parts

When replacement parts are required for the Yamaha FSI-E, it is advisable to deal direct with a recognised Yamaha agent or with the area distributor. They are better placed to supply the parts ex-stock and should have the technical experience that may not be available with other suppliers. When ordering parts, always quote the engine and frame numbers in full, since these will identify the model and its date of manufacture. Although the FSI-E is still comparatively new, it will sometimes help if the old part is presented when the replacement is ordered, to aid correct identification.

Always fit replacement parts of Yamaha manufacture and do not be tempted to use pattern parts, which sometimes have a price advantage. Although the pattern parts may appear similar they often give inferior service and may prove more expensive in the long run.

The engine number is stamped on the left hand crankcase immediately in front of the clutch cable. The frame number is stamped on the right hand side of the steering head.

Some of the more expendable parts such as spark plugs, bulbs, tyres, oils and greases etc., can be obtained from accessory shops and motor factors, who have convenient opening hours, charge lower prices and can often be found not far from home. It is also possible to obtain parts on a Mail Order basis from a number of specialists who advertise regularly in the motor cycle magazines.

Engine No. location

Frame number location

Routine maintenance

Refer to Chapter 7 for information relating to the 1975 on models

Periodic routine maintenance is a continuous process that commences immediately the machine is used. It must be carried out at specified mileage recordings or on a calendar basis if the machine is not used frequently, whichever falls soonest. Maintenance should be regarded as an insurance policy, to help keep the machine in the peak of condition and to ensure long, trouble-free service. It has the additional benefit of giving early warning of any faults that may develop and will act as a regular safety check, to the obvious advantage of both rider and machine alike.

The various maintenance tasks are described under their respective mileage and calendar headings. Accompanying diagrams are provided, where necessary. It should be remembered that the interval between the various maintenance tasks serves only as a guide. As the machine gets older or is used under particularly adverse conditions, it would be advisable to reduce the period between each check.

Some of the tasks are described in detail, where they are not mentioned fully as a routine maintenance item in the text. If a specific item is mentioned but not described in detail, it will be covered fully in the appropriate Chapter. No special tools are required for the normal routine maintenance tasks. The tools contained in the kit supplied with every new machine will prove adequate for each task or if they are not available, the tools found in the average household should suffice.

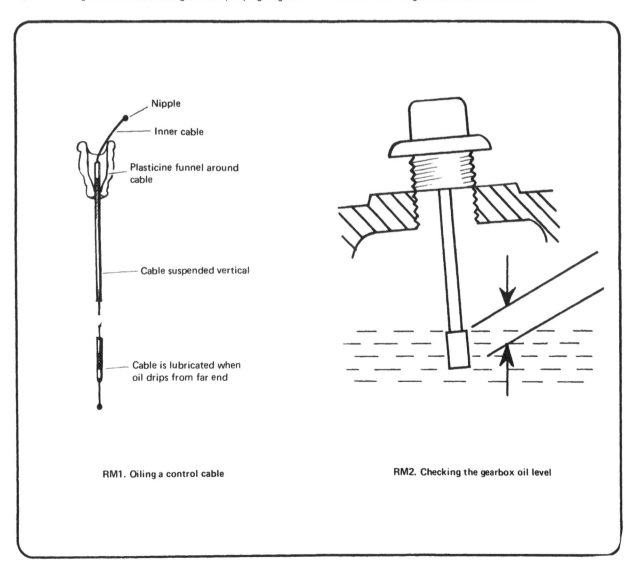

Nipple

Inner cable

Plasticine funnel around cable

Cable suspended vertical

Cable is lubricated when oil drips from far end

RM1. Oiling a control cable

RM2. Checking the gearbox oil level

Weekly or every 300 miles (500 km)

Check the tyre pressures. Always check with the tyres cold, using a pressure gauge known to be accurate.

Check the transmission oil level and top up if necessary. If the oil level is correct it will show on the dipstick but the dipstick must not be screwed in when checking the level. Note that the engine must be fully warmed up first and then switched off, also that the machine must be standing upright on its wheels.

Check and adjust the brakes, and oil the cable and lever pivots.

Check the acid level in the battery and top up with distilled water if necessary.

Check and, if necessary, adjust the tension of the drive chain. Make sure the chain is well lubricated.

Monthly or every 1000 miles (1600 km)

Complete all the checks listed in the weekly/300 mile service, and the following items:

Check the spark plug gap. If the electrodes are wearing thin, or if the outer electrode has to be bent excessively to restore the gap, fit a new plug.

Change the transmission oil.

Clean the carburettor, fuel tap and feed pipe. If necessary, re-adjust the slow running speed. Clean also the air filter.

Remove, clean and lubricate the drive chain.

Check the contact breaker gap and adjust if necessary.

Check lighting system.

Check the clutch adjustment. The clutch is adjusted correctly when there is 2 – 3 mm (0.08 – 0.12 in) of free play in the cable, measured between the butt end of the handlebar lever and its clamp, and when the clutch is operating correctly with no signs of slip or drag.

Normal adjustment is made at the cable lower adjuster, reserving the lever clamp adjuster, where fitted, for quick roadside adjustments. If this is no longer possible, or if there are signs of slip or drag, the mechanism must be adjusted as follows.

Slacken their locknuts and screw both adjusters fully in to gain the maximum cable free play. Remove the black rubber plug from the crankcase left-hand cover to expose the clutch adjuster. Slacken the adjuster locknut and unscrew (anticlockwise) the adjuster screw to ensure that there is no pressure on it, then screw it in (clockwise) until it seats lightly; do not overtighten it as this is the point where the mechanism is starting to lift the clutch pressure plate. Set the required free play by unscrewing (anticlockwise) the adjusting screw through 1/4 turn, then hold it while the locknut is tightened securely. Replace the plug and use the cable lower adjuster to set the specified free play at the handlebar. Apply a few drops of oil to the lower pivot, to the adjuster threads and to all exposed lengths of inner cable.

Six monthly or every 3000 miles (5000 km)

Complete all the checks under the weekly and monthly headings, then carry out the following additional tasks:

Decarbonise the engine and clean out the exhaust system.

Grease the centre stand pivot.

Grease the speedometer drive gears.

Lubricate the control cables, adjusting screws for front brake, rear brake and clutch.

Yearly or every 6000 miles (10,000 km)

Again complete all the checks listed under the weekly, monthly and six monthly headings, but only if they are not directly connected with the tasks listed below. Then complete the following:

Lubricate the felt wick of the contact breaker cam.

Check and, if necessary, replace both the final drive and pedalling gear chains. Check also the condition of the sprockets.

Adjust and lubricate the steering head bearings.

Dismantle both front and rear brake assemblies and examine the brake shoe linings. Replace the brake shoes if the linings are thin or if the leverage of the brake operating arm is reduced. Repack the bearings with grease.

Safety first!

Professional motor mechanics are trained in safe working procedures. However enthusiastic you may be about getting on with the job in hand, do take the time to ensure that your safety is not put at risk. A moment's lack of attention can result in an accident, as can failure to observe certain elementary precautions.

There will always be new ways of having accidents, and the following points do not pretend to be a comprehensive list of all dangers; they are intended rather to make you aware of the risks and to encourage a safety-conscious approach to all work you carry out on your vehicle.

Essential DOs and DON'Ts

DON'T start the engine without first ascertaining that the transmission is in neutral.

DON'T suddenly remove the filler cap from a hot cooling system – cover it with a cloth and release the pressure gradually first, or you may get scalded by escaping coolant.

DON'T attempt to drain oil until you are sure it has cooled sufficiently to avoid scalding you.

DON'T grasp any part of the engine, exhaust or silencer without first ascertaining that it is sufficiently cool to avoid burning you.

DON'T allow brake fluid or antifreeze to contact the machine's paintwork or plastic components.

DON'T syphon toxic liquids such as fuel, brake fluid or antifreeze by mouth, or allow them to remain on your skin.

DON'T inhale dust – it may be injurious to health (see *Asbestos* heading).

DON'T allow any spilt oil or grease to remain on the floor – wipe it up straight away, before someone slips on it.

DON'T use ill-fitting spanners or other tools which may slip and cause injury.

DON'T attempt to lift a heavy component which may be beyond your capability – get assistance.

DON'T rush to finish a job, or take unverified short cuts.

DON'T allow children or animals in or around an unattended vehicle.

DON'T inflate a tyre to a pressure above the recommended maximum. Apart from overstressing the carcase and wheel rim, in extreme cases the tyre may blow off forcibly.

DO ensure that the machine is supported securely at all times. This is especially important when the machine is blocked up to aid wheel or fork removal.

DO take care when attempting to slacken a stubborn nut or bolt. It is generally better to pull on a spanner, rather than push, so that if slippage occurs you fall away from the machine rather than on to it.

DO wear eye protection when using power tools such as drill, sander, bench grinder etc.

DO use a barrier cream on your hands prior to undertaking dirty jobs – it will protect your skin from infection as well as making the dirt easier to remove afterwards; but make sure your hands aren't left slippery. Note that long-term contact with used engine oil can be a health hazard.

DO keep loose clothing (cuffs, tie etc) and long hair well out of the way of moving mechanical parts.

DO remove rings, wristwatch etc, before working on the vehicle – especially the electrical system.

DO keep your work area tidy – it is only too easy to fall over articles left lying around.

DO exercise caution when compressing springs for removal or installation. Ensure that the tension is applied and released in a controlled manner, using suitable tools which preclude the possibility of the spring escaping violently.

DO ensure that any lifting tackle used has a safe working load rating adequate for the job.

DO get someone to check periodically that all is well, when working alone on the vehicle.

DO carry out work in a logical sequence and check that everything is correctly assembled and tightened afterwards.

DO remember that your vehicle's safety affects that of yourself and others. If in doubt on any point, get specialist advice.

IF, in spite of following these precautions, you are unfortunate enough to injure yourself, seek medical attention as soon as possible.

Asbestos

Certain friction, insulating, sealing, and other products – such as brake linings, clutch linings, gaskets, etc – contain asbestos. *Extreme care must be taken to avoid inhalation of dust from such products since it is hazardous to health.* If in doubt, assume that they *do* contain asbestos.

Fire

Remember at all times that petrol (gasoline) is highly flammable. Never smoke, or have any kind of naked flame around, when working on the vehicle. But the risk does not end there – a spark caused by an electrical short-circuit, by two metal surfaces contacting each other, by careless use of tools, or even by static electricity built up in your body under certain conditions, can ignite petrol vapour, which in a confined space is highly explosive.

Always disconnect the battery earth (ground) terminal before working on any part of the fuel or electrical system, and never risk spilling fuel on to a hot engine or exhaust.

It is recommended that a fire extinguisher of a type suitable for fuel and electrical fires is kept handy in the garage or workplace at all times. Never try to extinguish a fuel or electrical fire with water.

Note: *Any reference to a 'torch' appearing in this manual should always be taken to mean a hand-held battery-operated electric lamp or flashlight. It does **not** mean a welding/gas torch or blowlamp.*

Fumes

Certain fumes are highly toxic and can quickly cause unconsciousness and even death if inhaled to any extent. Petrol (gasoline) vapour comes into this category, as do the vapours from certain solvents such as trichloroethylene. Any draining or pouring of such volatile fluids should be done in a well ventilated area.

When using cleaning fluids and solvents, read the instructions carefully. Never use materials from unmarked containers – they may give off poisonous vapours.

Never run the engine of a motor vehicle in an enclosed space such as a garage. Exhaust fumes contain carbon monoxide which is extremely poisonous; if you need to run the engine, always do so in the open air or at least have the rear of the vehicle outside the workplace.

The battery

Never cause a spark, or allow a naked light, near the vehicle's battery. It will normally be giving off a certain amount of hydrogen gas, which is highly explosive.

Always disconnect the battery earth (ground) terminal before working on the fuel or electrical systems.

If possible, loosen the filler plugs or cover when charging the battery from an external source. Do not charge at an excessive rate or the battery may burst.

Take care when topping up and when carrying the battery. The acid electrolyte, even when diluted, is very corrosive and should not be allowed to contact the eyes or skin.

If you ever need to prepare electrolyte yourself, always add the acid slowly to the water, and never the other way round. Protect against splashes by wearing rubber gloves and goggles.

Mains electricity and electrical equipment

When using an electric power tool, inspection light etc, always ensure that the appliance is correctly connected to its plug and that, where necessary, it is properly earthed (grounded). Do not use such appliances in damp conditions and, again, beware of creating a spark or applying excessive heat in the vicinity of fuel or fuel vapour. Also ensure that the appliances meet the relevant national safety standards.

Ignition HT voltage

A severe electric shock can result from touching certain parts of the ignition system, such as the HT leads, when the engine is running or being cranked, particularly if components are damp or the insulation is defective. Where an electronic ignition system is fitted, the HT voltage is much higher and could prove fatal.

Lubrication chart

1 Remove, clean and lubricate final drive chain; check tension
2 Check acid level of battery
3 Check points gap and lubricate wick of contact breaker cam
4 Check oil level regularly
5 Grease speedometer drive
6 Clean and adjust sparking plug
7 Grease steering head bearings
8 Lubricate control cables

Recommended lubricants

Fuel	Unleaded, or leaded two-star (regular)
Engine:	
Petroil models (1972 to 1977)	Good quality self-mixing two-stroke oil, 20:1 mixing ratio, ie 0.4 (2/5) pint/227 cc oil to 1 gal petrol, 50 cc oil to 1 lit petrol
Autolube models (1977 on)	Good quality air-cooled 2-stroke engine oil
Transmission (gearbox)	SAE10W/30 SE engine oil
Front forks	SAE10W/30 SE engine oil or SAE 10 fork oil
Final drive chain	Commercial chain lubricant
Bearings and other greasing points	Multi-purpose high melting-point lithium-based grease
Control cables and general lubrication	Light machine oil

Dimensions and weights

Dimensions and weights:

Overall length	69.1 in (1755 mm)
Overall width	21.9 in (555 mm)
Overall height	36.8 in (935 mm)
Wheelbase	45.7 in (1160 mm)
Ground clearance (unloaded)	5.3 in (135 mm)
Dry weight	154 lb (70 kg)

Chapter 1 Engine clutch and gearbox

Refer to Chapter 7 for information relating to the 1975 on models

Contents

Specifications

Engine:

Type	Two-stroke, with loop scavenging and rotary disc inlet valve
Cylinder head	Aluminium alloy, deeply finned
Cylinder barrel	Cast iron, deeply finned
Bore	40 mm
Stroke	39.7 mm
Capacity	49 cc
Bhp	4.8 bhp at 7,000 rpm
Compression ratio	7 : 1
Petrol/oil ratio	20 : 1

Piston:

Type	Flat top, with transfer port cutaways in base of skirt. Pegged, to retain piston rings
Oversizes available	+ 0.010 in. (0.25 mm) and + 0.020 in. (0.50 mm)

Piston rings:

Type	Two only, both compression. Profiled ends to locate with piston pegs, one chrome plated, one parkerised
End gap	0.006 in. (0.15 mm) — 0.014 in. (0.35 mm)

Cylinder barrel:
Bore ... ...   40 mm
Limits ... ...   Maximum permissible ovality 0.002 in. (0.05 mm)

Crankshaft:
Type ... ...   Steel, two bearing with caged roller big and small ends

Torque wrench settings:

Stud size (mm)	ft lb	Stud size (mm)	ft lb
6	7.5	12	29.0 – 33.0
7	11.25	14	33.0 – 37.5
8	15.0	17	41.5 – 50.0
10	25.0 – 29.0		

Gear ratios:
Top 1.038
Third 1.304
Second 1.889
Bottom 3.077

Clutch ... ...   Wet, multi disc
Primary drive ... ...   Gear
Primary drive ratio 3.895 (74/19)
Final drive ... ...   Chain
Final drive ratio 2.785 (39/14)

1 General description

The engine fitted to the Yamaha FS1-E is of the two stroke type, with a rotary disc inlet valve, working on the 'loop scavenging' principle. A flat top piston is used, fitted with pegs to retain the piston rings in a set location so that they cannot rotate and permit the ends to become trapped and broken in the ports. Cutaways in the base of the piston skirt facilitate the opening and closing of the transfer ports.

Lubrication is effected by a petrol/oil mixture which must be premixed in the fuel tank in the recommended proportions of 20 parts of petrol to one part of oil. The system operates on the 'total loss' principle whereby all excess oil is expelled via the exhaust system. Because the two-stroke utilises crankcase compression before the incoming mixture is transferred to the cylinder for ignition, the big end, main bearings, small end and piston are fully lubricated by the oil content of the incoming charge.

The gearbox and clutch depend on separate lubrication and run in oil in a separate compartment of the crankcase assembly.

The engine and gearbox are of unit construction hence when the crankcases are split the crankshaft and gearbox internals are exposed.

2 Operations with engine in frame

It is not necessary to remove the engine unit from the frame unless the crankshaft assembly and/or the gearbox internals require attention. Most operations can be accomplished with the engine in place, such as removal and replacement of:

1 Cylinder head
2 Cylinder barrel and piston
3 Flywheel generator
4 Clutch assembly
5 Contact breaker assembly

When several operations need to be undertaken simultaneously, it will probably be advantageous to remove the complete engine unit from the frame, an operation that should take approximately fifteen minutes. This will give the advantage of better access and more working space.

3 Operations with engine removed

1 Removal and replacement of the main bearings.
2 Removal and replacement of the crankshaft assembly.
3 Removal and replacement of the gear cluster, selectors and gearbox main bearings.

4 Method of engine/gearbox removal

As described previously, the engine and gearbox are built in unit and it is necessary to remove the unit complete in order to gain access to either component. Separation is accomplished after the engine unit has been removed and refitting cannot take place until the crankcases have been reassembled. When the crankcases are separated the gearbox internals will also be exposed.

5 Removing the engine/gear unit

1 Place the machine on the centre stand and make sure it is standing firmly on level ground.
2 Turn off the petrol tap.
3 Remove the side covers, the right hand one contains the tool kit, the left hand one exposes the battery.
4 Disconnect the battery and remove it from the machine.
5 Remove the cotter pin from the left hand pedal crank. The cotter pin may be very tight so care should be taken not to damage the thread.
6 Remove the three screws in the pedal chain cover and let it drop down the pedal crank.
7 Remove the circlip retaining the pedal drive sprocket and slide the sprocket and crank off together, followed by the spring and drive dog. Pull the other crank, with the spindle, from the right hand side of the machine.
8 Ensure that the machine is in neutral, then remove the gearchange lever bolt completely which will allow the lever to slide off its shaft.
9 Remove the four screws and the left hand cover to reveal the generator. It will lift away with the clutch cable attached. There is no need to separate the cable.
10 Disconnect the generator leads in the battery compartment.
11 Remove the final drive chain spring link and pull the chain off the engine sprocket.

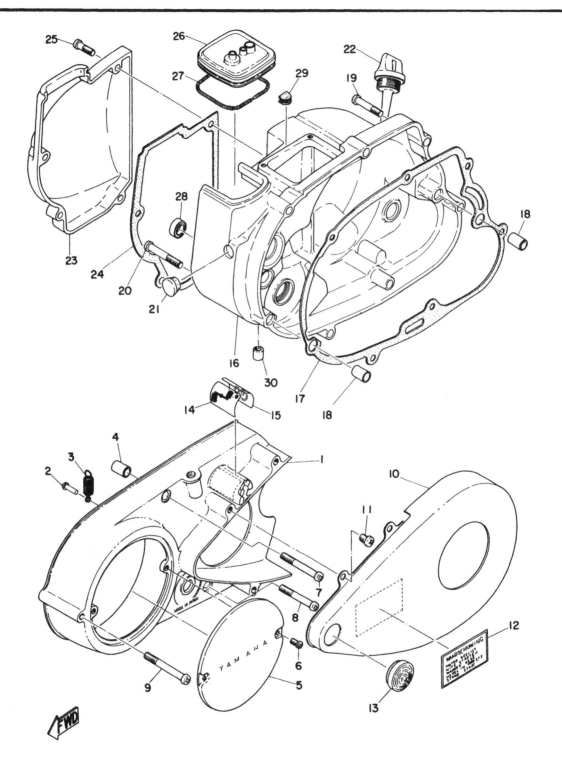

Fig. 1.1. Crankcase covers

1	Left-hand crankcase cover	8	Crankcase cover screw
2	Spring hook	9	Crankcase cover screw
3	Return spring (clutch operating arm)	10	Pedalling chain cover
4	Dowel pin - 2 off	11	Panhead screw for chain cover - 3 off
5	Generator cover	12	Transfer for chain cover
6	Panhead screws for generator cover - 2 off	13	Chain cover cap
		14	Instruction transfer 1
7	Crankcase cover screw - 2 off	15	Instruction transfer 2
		16	Right-hand crankcase cover

17	Gasket for right-hand cover	24	Gasket for carburettor cover
18	Dowel pin - 2 off	25	Panhead screw for carburettor cover - 4 off
19	Panhead screw for right-hand cover - 6 off	26	Boot for carburettor cover
20	Panhead screw for right-hand cover	27	Spring sealing band
21	Grommet plug	28	Blind plug
22	Oil level plug	29	Grommet
23	Carburettor cover	30	Drain tube

12 From the right hand side of the machine, remove the air filter end caps and element.

13 Remove the four screws and right hand cover to reveal the carburettor.

14 Slide the rubber boot and spring retainer up the control cables and pull the petrol pipe off.

15 Remove the plastic bung from the front of the carburettor enclosure, insert a screwdriver in the hole and slacken off the carburettor clamp ring. Pull the carburettor off its stub and tie it up out of the way.

16 Remove the kickstarter bolt completely and take off the kickstarter lever.

17 Undo the exhaust ring nut and take off the swinging arm nut to release the silencer. Gently pull the exhaust system from the machine and allow the tension of the brake return spring to be relieved. Unhook the spring and the exhaust system can be taken clear.

18 Remove the two bolts holding the air cleaner case and move the case to enable the four screws holding the top cover to be removed. The air cleaner case and the top cover through which the fuel pipe passes can then also be tied up out of the way.

19 Remove the spark plug, plug cap and the rear brake light switch.

20 Remove the top engine mounting bolt and allow the engine the pivot on the bottom bolt so that it rests on the ground.

21 Remove the bottom engine bolt and pull the engine clear of the machine.

6 Dismantling the engine, clutch and gearbox - general

1 Before commencing work on the engine unit, the external surfaces should be cleaned thoroughly. A motorcycle engine has very little protection from road grit and other foreign matter, which will find its way into the dismantled engine if this simple precaution is not observed. One of the proprietary cleaning compounds such as Gunk can be used to good effect, particularly if the compound is allowed to work into the film of oil and grease before it is washed away. When washing down, make sure that water cannot enter the carburettor or the electrical system, particularly if these parts have been exposed.

2 Never use undue force to remove any stubborn part, unless mention is made of this requirement. There is invariably good reason why a part is difficult to remove, often because the dismantling operation has been tackled in the wrong sequence. Dismantling will be made easier if a simple engine stand is constructed that will correspond with the engine mounting points. This arrangement will permit the complete unit to be clamped rigidly to the workbench, leaving both hands free.

7 Generator - removal

1 The generator rotor and stator may be removed with the engine in the frame or removed from it. In the former case, it will be necessary to carry out first the preliminary dismantling operations described in the first ten paragraphs of Section 5 of this Chapter.

2 Before the rotor retaining nut can be removed the crankshaft must be locked to prevent rotation. If the engine is in the frame, this can be achieved by selecting top gear and applying hard the rear brake, thus locking the crankshaft via the transmission. If the engine is removed from the frame a holding tool (such as a strap wrench) can be applied to the rotor, or if the cylinder head, barrel and piston have been removed, a close-fitting metal bar can be passed through the connecting rod small-end eye and rested on two wooden blocks placed across the crankcase mouth. Remove the nut and its spring and plain washers.

3 Remove the rotor using only the Yamaha service tool 90890-01148 or a pattern version of it available from most good motorcycle dealers. If necessary, take the machine to a Yamaha Service Agent for the rotor to be removed. If all else fails, an alternative method can be employed, using equipment similar to that shown in the accompanying photograph, but great care must be taken not to damage the crankshaft end or the rotor itself. Never attempt to remove the rotor using levers.

4 Displace the Woodruff key from the crankshaft keyway, disconnect the neutral indicator switch wire, then use an impact driver to remove the two stator plate mounting screws and withdraw the stator plate.

8 Dismantling the engine unit - removing the cylinder head and cylinder barrel and piston

1 Unscrew the four nuts which retain the cylinder head in position and remove them together with their washers. The cylinder head can now be lifted off the holding down studs.

2 Slide the cylinder barrel up the holding down studs, taking care to support the piston when it falls clear of the cylinder bore. If only a limited amount of dismantling is being undertaken, it is advisable to pad the crankcase mouth with clean rag as soon as the cylinder barrel is raised, otherwise particles of broken piston ring may fall into the crankcase and necessitate further dismantling to retrieve them.

3 Remove the two circlips from the piston, using a pair of long nosed pliers. The gudgeon pin can now be pushed out of position, allowing the piston complete with rings to be removed from the connecting rod.

5.4 Disconnect battery and remove from machine

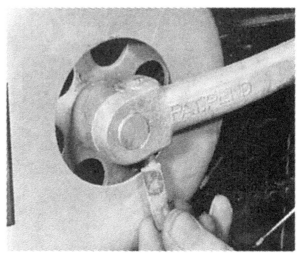

5.5 Drive cotter pin from left-hand pedal crank

5.7a Circlip retains pedal drive sprocket

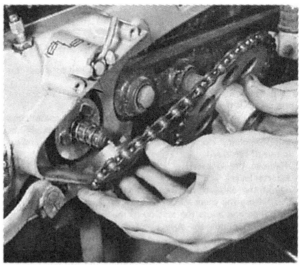

5.7b Lift off pedal crank, sprocket and chain in unison

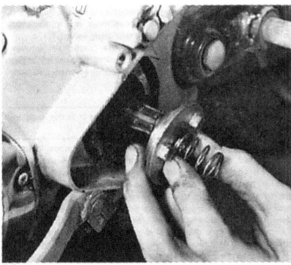

5.7c Follow by lifting off spring and drive dog

5.8 Remove bolt completely to withdraw gear change lever

5.9 Generator cover is retained by four screws

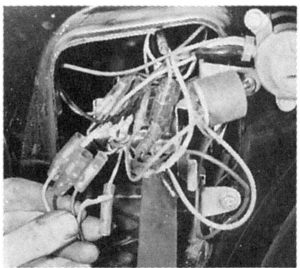

5.10 Generator leads are colour coded to make reconnection easy

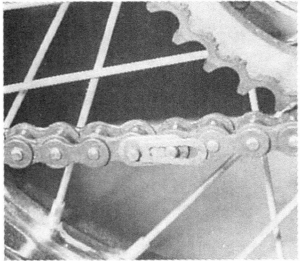

5.11 Remove spring link to separate final drive chain

5.12 Air cleaner element pulls out of casing

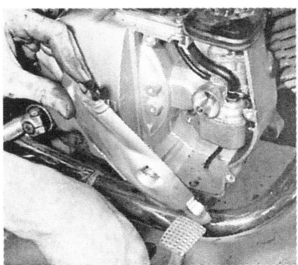

5.13 Remove right-hand cover for access to carburettor

5.14 Raise rubber boot and lift off petrol feed pipe

5.15a Access to carburettor clamp is through hole in cover

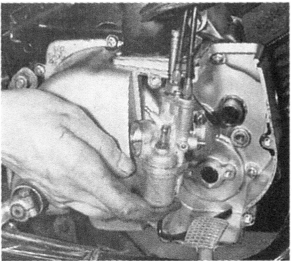

5.15b Carburettor will pull off inlet stub

5.16 Remove kickstarter bolt completely to free crank

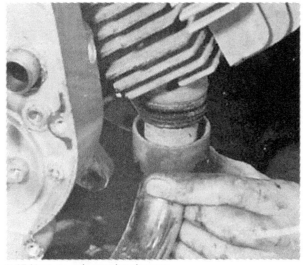

5.17 Unscrew exhaust pipe ring nut

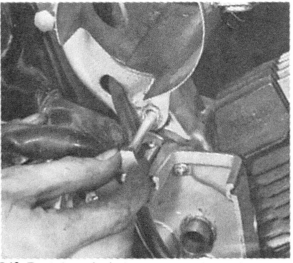

5.18a Two bolts retain air cleaner case to engine and frame

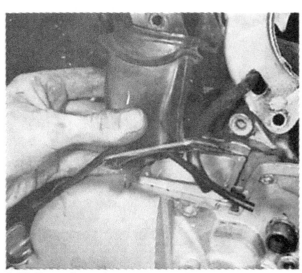

5.18b Four bolts retain top cover to carburettor compartment

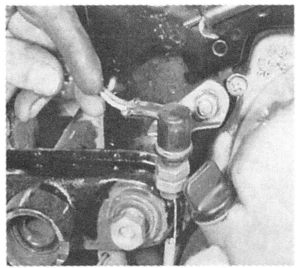

5.19 Unhook spring and remove stop lamp switch

5.20 Removal of top engine bolt allows engine unit to pivot downwards

7.1 Lock engine to remove nut and washer

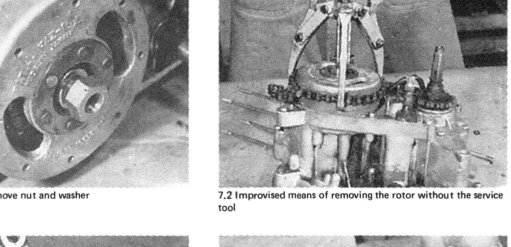

7.2 Improvised means of removing the rotor without the service tool

7.3 Stator plate is retained by two cross head screws

8.1 Cylinder head is retained by four nuts

4 If the gudgeon pin is a tight fit, warm the piston by placing a rag soaked in hot water on the crown. Never drift the gudgeon pin out of position without supporting the piston, otherwise there is risk of bending the connecting rod. Throw away the circlips; they should never be re-used.

5 Note that the piston crown is marked with an arrow and when reassembling the piston this arrow should point down (toward the exhaust port).

6 The small end is a caged roller assembly and will slide easily out of the connecting rod.

9 Dismantling the engine unit - final drive sprocket removal

1 Remove the circlip from the gearbox layshaft and flatten the raised tab of the sprocket retaining nut lock washer.

2 If the engine unit is in the frame, apply hard the rear brake to lock the sprocket while the nut is removed. If the engine unit has been removed from the frame, use a chain wrench as shown in the accompanying photograph; a substitute chain wrench can be fabricated easily using the machine's own chain.

3 Remove the nut, displace the two collets and slide off the threaded collar, followed by the lock washer. The sprocket can then be slid off the layshaft splines.

8.2 Cylinder barrel will slide up holding down studs

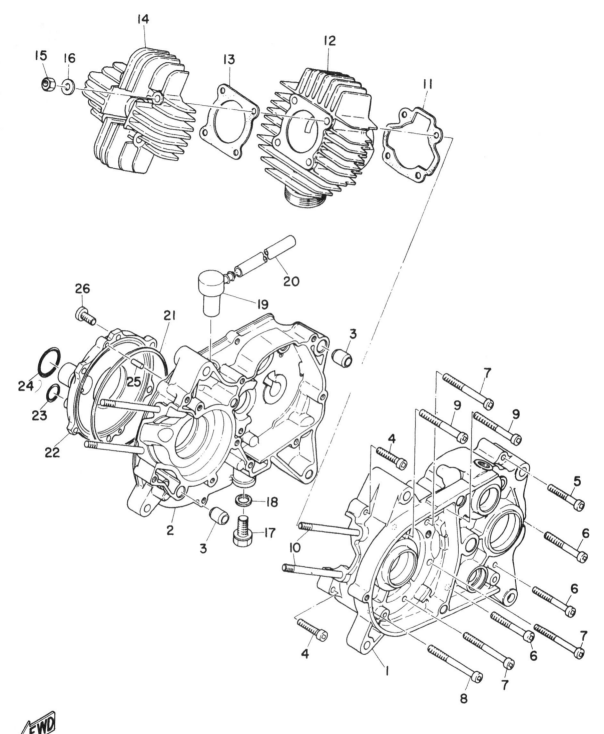

Fig. 1.2. Crankcase and cylinder assembly

1 Left-hand crankcase
2 Right-hand crankcase
3 Dowel pin - 2 off
4 Panhead screw for left-hand crankcase - 2 off
5 Panhead screw for left-hand crankcase
6 Panhead screw for left-hand crankcase - 3 off
7 Panhead screw for left-hand crankcase - 3 off
8 Panhead screw for left-hand crankcase
9 Panhead screw for left-hand crankcase - 2 off
10 Cylinder holding down stud - 4 off
11 Cylinder base gasket
12 Cylinder barrel
13 Cylinder head gasket
14 Cylinder head
15 Cylinder head nut - 4 off
16 Plain washer - 4 off
17 Drain plug
18 Drain plug washer
19 Breather assembly
20 Breather pipe
21 'O' ring seal for disc valve cover
22 Disc valve cover
23 'O' ring seal
24 'O' ring seal for carburettor stub
25 Dowel pin - 2 off
26 Panhead screw for disc valve cover - 6 off

8.3 Long nose pliers aid removal of circlips

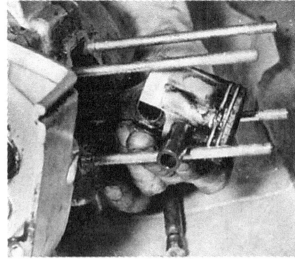

8.4 Heat piston if gudgeon pin is a tight fit

8.6 Small end bearing is a caged needle roller assembly

9.1 Remove circlip from gearbox layshaft

9.2 Use of chain wrench to lock sprocket whilst slackening nut

9.3a Remove two collets and screwed sleeve to free sprocket ...

9.3b ...which will then lift off layshaft

10.2 Right-hand cover is retained by seven screws

10 Dismantling the engine unit - clutch dismantling and removal

1 Drain the engine oil.
2 Remove the seven screws holding the right hand cover, noting that the long one is fitted by the carburettor stub.
3 Remove the cover with the engine over a small tray to catch the oil left in the engine. Note that there are two O rings, one on the carburettor stub and one adjacent to it which may stick to the cover; these should be removed to avoid loss or damage.
4 Remove the four screws and clutch pressure springs. The clutch pressure plate can then be removed to expose the pushrod.
5 The pushrod is in two pieces with a very small ball bearing between them so after pulling out the first pushrod, tip the engine and catch the small ball as it rolls out. The second pushrod can be pulled out from the opposite side.
6 Remove the three clutch plates, one aluminium, two fibre.
7 Prise the tab washer away from the clutch centre nut, lock the clutch centre with a suitable piece of shaped metal and undo the nut.
8 The tab washer and clutch centre can then be removed, followed by a thrust washer (where fitted) and a thrust plate.
9 The clutch drum can then be pulled off with a slight clockwise turn to free the helical gears.

10.4a Remove the four screws and springs to free pressure plate

10.4b Lift out first portion of clutch push rod

10.7 Use piece of shaped metal to lock clutch centre whilst slackening nut

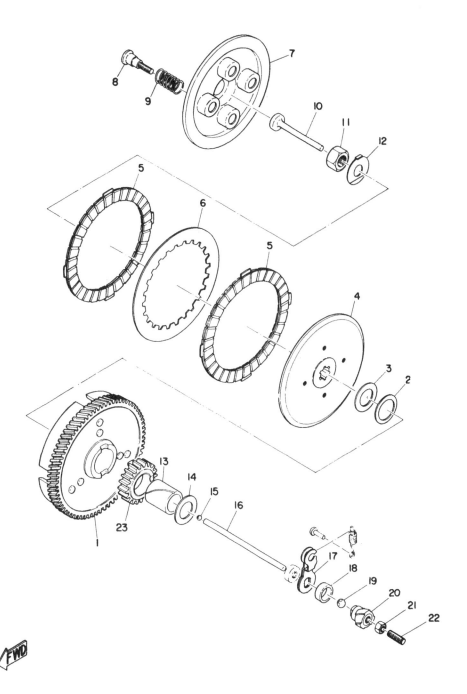

Fig. 1.3. Clutch

1	Clutch outer drum	6	Plain plate	12	Tab washer	18	Oil seal
2	Thrust washer (early models only)	7	Pressure plate	13	Spacer	19	Ball bearing (¼ inch)
3	Thrust plate	8	Clutch spring screw - 4 off	14	Thrust plate	20	Quick thread worm
4	Clutch centre	9	Clutch spring - 4 off	15	Ball bearing (3/16 inch)	21	Adjuster nut
5	Friction plate - 2 off	10	Push rod 1	16	Push rod 2	22	Adjusting screw
		11	Clutch centre retaining nut	17	Actuating lever	23	Kickstarter pinion

10.8 Note thrust washer behind clutch centre

10.9 Lift off with clockwise twist to disengage helical gears

11.2 Circlip retains idler pinion

11.1 Unhook kickstarter return spring and pull assembly from crankcase

11.3 Clutch pinion and sleeve assembly will then lift off

11 Dismantling the engine - kickstarter mechanism

1 When the clutch has been removed the kickstarter shaft assembly can be pulled from the crankcases, when the spring has been unhooked.

2 Remove the circlip holding the idler pinion and pull off the plain washer, the pinion, the crinkle washer and a plain washer.

3 The clutch gear, sleeve and thrust washer can then be taken off.

4 To change the kickstarter return spring, remove the circlip and slotted spring register collar. The spring can then be removed from the shaft.

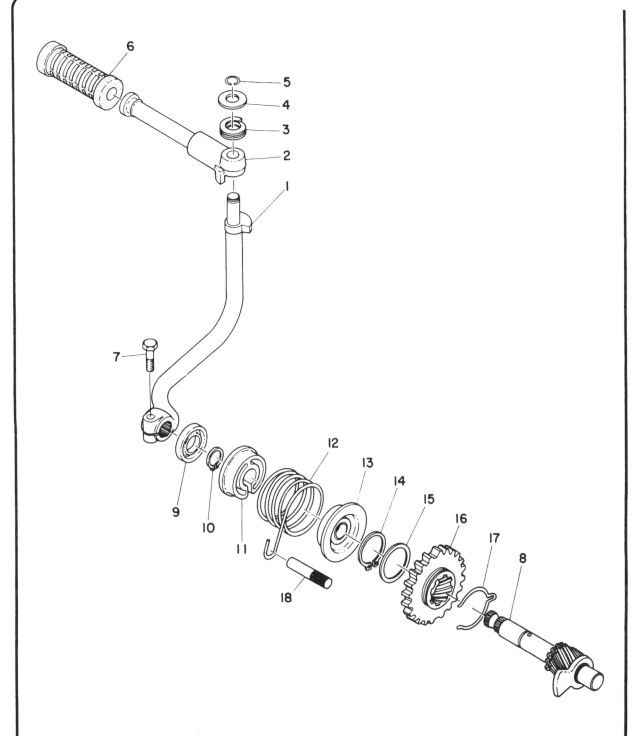

Fig. 1.4. Kickstarter assembly

1 Kickstarter crank
2 Kickstarter lever
3 Kickstarter crank spring
4 Kickstarter lever washer
5 Kickstarter lever clip

6 Kickstarter lever rubber
7 Bolt for kickstarter crank
8 Kickstarter spindle assembly
9 Oil seal
10 Circlip

11 Kickstarter return
spring collar
12 Kickstarter return spring
13 Kickstarter return
spring guide

14 Circlip
15 Shim
16 Kickstarter drive pinion
(26 teeth)
17 Kickstarter assembly clip
18 Kickstarter spring anchor

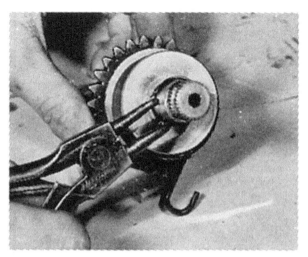

11.4a Remove circlip, followed by ...

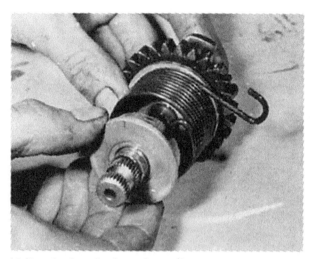

11.4b ... the slotted spring register collar

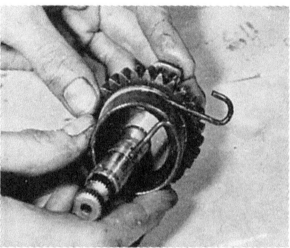

11.4c End of spring engages with hole in spindle ...

11.4d ... seats on another shouldered collar

12 Dismantling the engine - gearchange mechanism and index arm

1   When the clutch has been removed, the circlip and washer on the gearchange shaft on the other side of the engine has to be removed, which, after the pawl arm has been lifted clear of the gearchange cam, will allow the shaft assembly to be withdrawn. The pawl spring and the gearchange return spring should be checked and renewed if necessary.
2   Remove the shouldered bolt holding the index arm and unhook the spring which allows the arm to be removed.

13 Dismantling the engine - disc valve removal

1   When the gearchange mechanism has been removed, the mainshaft nut and belleville washer are taken off which allows the helical gear and spacer to be slid off its splines. Lock the engine to aid removal by inserting a metal rod through the eye of the connecting rod, so that it rests across the crankcase mouth.
2   Undo the six screws and remove the cover plate to expose the valve disc. Note that the pin on the mainshaft is in line with two indentations on the disc and must be lined up for reassembly.
3   Carefully remove the disc as it is made of fibre and is easily broken. Care should be taken when cleaning the disc as some cleaning fluids can destroy the disc. Wiping with a clean rag is the best recommendation.
4   The drive collar can then be slid off the crankshaft and the pin in the crankshaft pulled or tapped out.

14 Dismantling the engine - separating the crankcases

1   There are twelve screws holding the crankcase halves together. There are six different lengths of screw so a note should be made as to which screw fits in which hole.
2   Before trying to separate the crankcases, ensure that the valve disc pin and the rotor Woodruff key have been removed.
3   The right hand crankcase should lift off but light tapping with a rawhide mallet may be necessary.
4   Never use the point of a screwdriver to aid the separation of the crankcases. A leaktight crankcase is an essential requirement of any two-stroke engine. If the crankcase joint is damaged in any way, air will be admitted, which will dilute the incoming mixture whilst it is under crankcase compression. This will cause poor starting and indifferent running, mainly due to the weakened mixture that results.
5   Note that there are two locating dowels fitted in the crankcases.

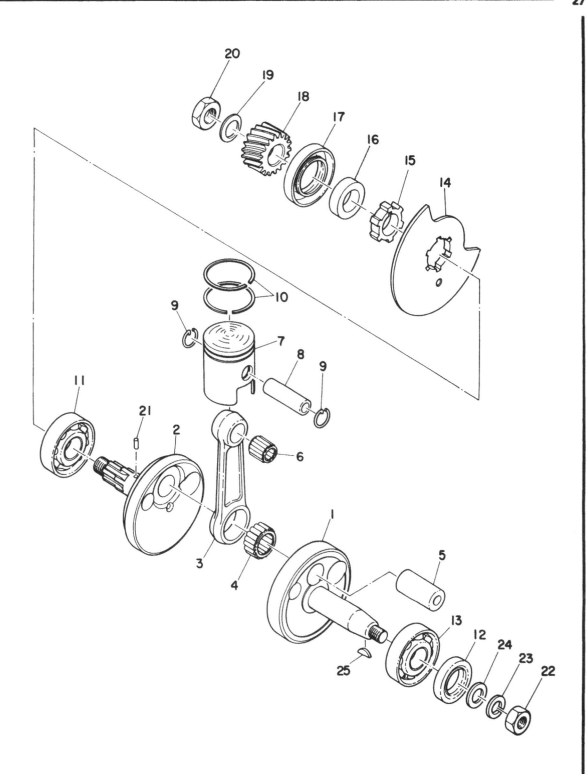

Fig. 1.5. Crankshaft, piston and disc valve

1	Left-hand crank	7	Piston (standard and	13	Left-hand main bearing
2	Right-hand crank		two oversizes available)	14	Disc valve
3	Connecting rod	8	Gudgeon pin	15	Disc valve collar
4	Big-end bearing	9	Circlip - 2 off	16	Distance piece
5	Crankpin	10	Piston ring set	17	Right-hand oil seal
6	Small end bearing	11	Right-hand main bearing	18	Primary drive pinion
		12	Left-hand oil seal	19	Belville washer

20 Pinion securing nut
21 Dowel pin
22 Nut
23 Spring washer
24 Palin washer
25 Woodruff key

12.1a Remove the circlip and washer from the gear change shaft

12.1b Lift pawl from gear change cam and withdraw shaft assembly

12.2 Remove shouldered bolt to release indexing arm and spring

13.1 Lock engine again to remove crankshaft nut and drive pinion

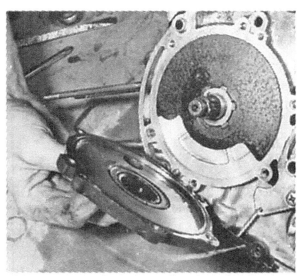

13.2 Remove cover plate to expose disc valve

13.3 Ease disc out with care

13.4 Drive collar will slide off crankshaft

15.2 The gearbox components after separating the crankcases

15 Dismantling the engine - removing the gearbox components

1   Remove the neutral indicator switch.
2   The gearbox shafts and selectors will slide out of the crankcase as a cluster but if difficulty is experienced check that the selector drum is in the neutral position.
3   If any damage or wear has occurred in the gearbox, the removal of the circlips on the gearshafts or selector shafts will allow the component concerned to be replaced. The gearchange drum is a built up assembly and should be renewed rather than repaired.
4   The final drive spacer will be left in the oil seal in the crankcase half and can be easily pulled out.

16 Dismantling the engine - removing the crankshaft assembly

Support the crankcase half and tap the end of the mainshaft with a rawhide mallet, to avoid damaging the thread, until the crankshaft assembly is free.

17 Crankshaft and gearbox main bearings - removal

1   The crankshaft assembly runs in two journal ball bearings

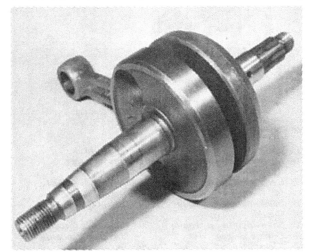

16.1 The complete crankshaft assembly

and the gearbox shafts in two journal ball bearings and two phosphor bronze bushes.
2   To remove the gearbox mainshaft journal bearing, the two screws and the bearing retaining plate must be withdrawn.
3   Before the bearings can be removed, it is necessary to warm the crankcases by applying a rag soaked in hot water. The bearings can then be tapped out. If the phosphor bronze bearings need renewing the new bearing can be used to push out the old one if a bolt and spacer arrangement is built up.

18 Oil seals - examination and replacement

1   Two stroke engines are particularly vulnerable to wear or damage which may occur to the oil seals. This is even so on the Yamaha engine. Apart from the oil leakage that will result, worn crankshaft seals will admit air to the crankcase and dilute the incoming mixture whilst it is under crankcase compression.
2   Early warning of failing crankshaft oil seals is given by difficulty in starting and a general reluctance for the engine to run smoothly.
3   The oil seals can be prised out of the crankcase but if removed new ones must be fitted.
4   The oil seals must be fitted with the manufacturer's marks and letters facing outward.
5   When fitting new oil seals, extreme care should be taken when they are inserted over a shaft. To prevent damage to the feather edge of the seal, grease both the shaft and the centre of the oil seal itself.

19 Crankshaft assembly - examination and renovation

1   Wash the complete crankshaft assembly with a petrol/paraffin mix to remove all surplus oil.
2   Replace the small end caged roller bearing and the gudgeon pin. Check for play between the gudgeon pin and the crankshaft. The play should not exceed 0.080 inch (2 mm).
3   If there is more play than is allowable the whole of the crankshaft assembly should be renewed as it is the connecting rod which has worn. To separate the flywheels is a highly specialised task, beyond the normal home garage facilities.

20 Cylinder barrel - examination and renovation

1   There will probably be a lip at the uppermost end of the cylinder barrel which marks the limit of travel of the top piston ring. The depth of the lip will give some indication of the

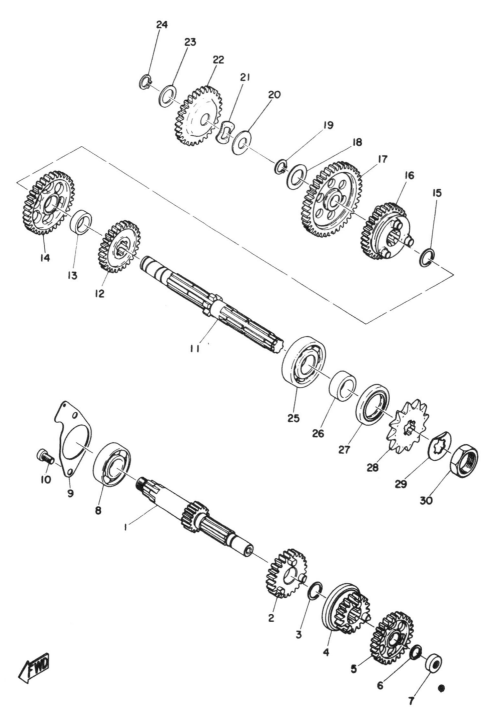

Fig. 1.6. Gearbox components

1	Mainshaft	8	Right-hand mainshaft	15	Clip
2	Mainshaft 3rd gear pinion		bearing	16	Layshaft 3rd gear pinion
	(23 teeth)	9	Bearing cover plate		(30 teeth)
3	Clip	10	Panhead screw for cover	17	Layshaft 1st gear pinion
4	Mainshaft 2nd gear pinion		plate - 2 off		(40 teeth)
	(18 teeth)	11	Layshaft	18	Shim - number as required
5	Mainshaft 4th gear pinion	12	Layshaft 4th gear pinion	19	Circlip
	(26 teeth)		(27 teeth)	20	Shim - number as required
6	Circlip	13	Distance collar	21	Crinkle washer
7	Push rod seal	14	Layshaft 2nd gear pinion	22	Kickstarter idler pinion
			(34 teeth)		

	(26 teeth)
23	Thrust washer
24	Circlip
25	Left-hand layshaft
	bearing
26	Distance collar
27	Oil seal
28	Final drive sprocket
	(14 teeth standard)
29	Tab washer
30	Sprocket retaining nut

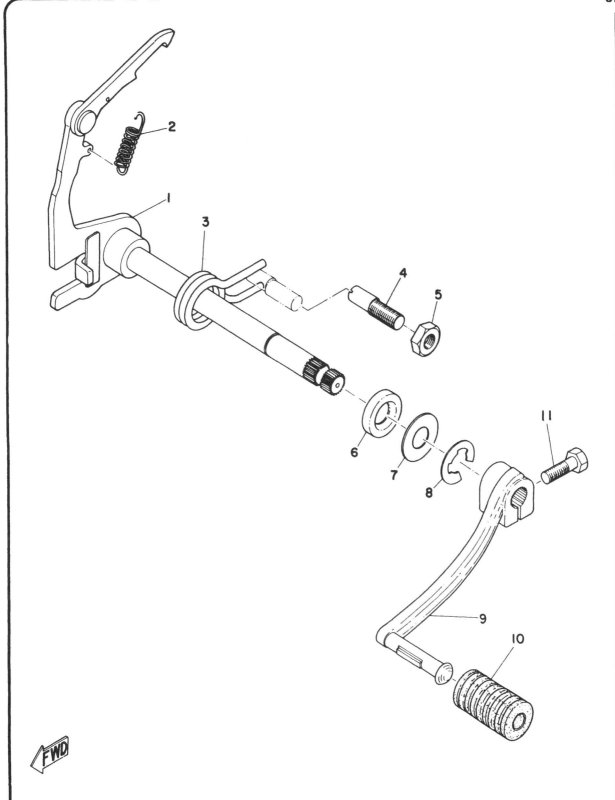

Fig. 1.7. Gear change mechanism

1 Gear change spindle
  assembly
2 Pawl spring

3 Gear change spindle
  return spring
4 Adjuster screw
5 Adjuster locknut

6 Oil seal
7 Washer
8 Circlip
9 Gear change pedal

10 Rubber for gear change
   pedal
11 Pedal retaining bolt

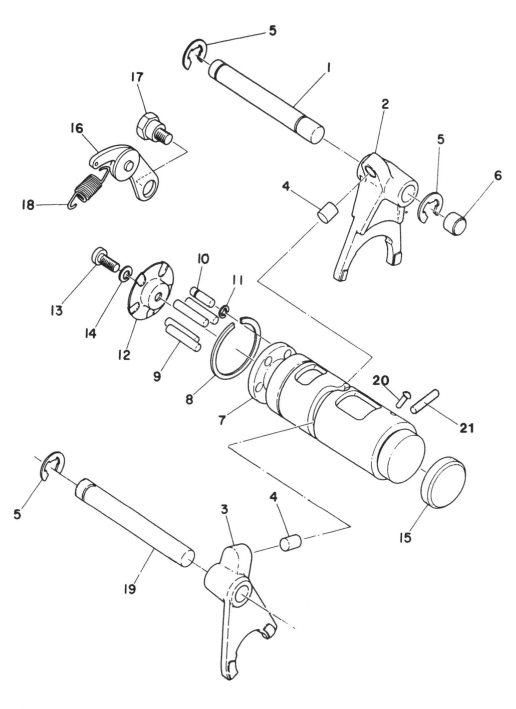

Fig. 1.8. Gear selector drum

1 Gear selector rod	5 Circlip - 3 off	11 Circlip	16 Index arm
2 Gear selector fork 1	6 Blind plug	12 Side plate	17 Shouldered bolt
3 Gear selector fork 2	7 Gear selector drum	13 Panhead screw for side	18 Index arm spring
4 Cam follower pin -	8 Circlip	plate	19 Gear selector rod
2 off	9 Dowel pin - 4 off	14 Spring washer	20 Neutral switch contact
	10 Locating pin	15 Blanking off plug	21 Dowel pin

amount of bore wear that has taken place, even though the amount of wear is not evenly distributed.

2   Remove the rings from the piston taking great care as they are brittle and very easily broken. There is more tendency for the rings to gum in their grooves in a two-stroke engine. Insert the piston in the bore so that it is positioned about ½ inch below the top of the bore. If it is possible to insert a 0.014 inch feeler gauge between the piston and the bore, a rebore and the fitting of an oversize piston is necessary.

3   Give the cylinder barrel a close visual inspection. If the surface of the bore is scored or grooved, indicative of an earlier seizure or a displaced circlip and gudgeon pin, a rebore is essential. Compression loss will have a very marked effect on performance.

4   Check that the outside of the cylinder barrel is clean and free from road dirt. Use a wire brush on the cooling fins if they are obstructed in any way. The engine will overheat badly if the cooling area is obstructed in any way. The application of matt cylinder black will help improve heat radiation.

5   Clean all carbon deposits from the exhaust ports and try and obtain a smooth finish in the ports without in any way enlarging them or altering their shape. The size and position of the ports predetermines the characteristics of the engine and unwarranted tampering can produce very adverse effects. An enlarged or re-profiled port does not necessarily guarantee an increase in performance.

21 Piston and piston rings - examination and renovation

1   Attention to the piston and piston rings can be overlooked if a rebore is necessary because new replacements will be fitted.

2   If a rebore is not considered necessary, the piston should be examined closely. Reject the piston if it is badly scored or if it is badly discoloured as the result of the exhaust gases by-passing the rings.

3   Remove all carbon from the piston crown and use metal polish to finish off. Carbon will not adhere so readily to a polished surface.

4   Check that the gudgeon pin bosses are not worn or the circlip grooves damaged. Check also the piston ring pegs, to make sure that none has worked loose.

5   The grooves in which the piston rings locate can also become enlarged in use. The clearance between the piston and the ring in the groove should not exceed 0.002 inch.

6   Piston ring wear can be checked by inserting the rings in the cylinder bore from the top and pushing them down about 1½ inches with the crown of the piston so that they rest square in the cylinder. If the end gap exceeds 0.014 inch the rings should be replaced.

7   Examine the working surface of the rings. If discoloured areas are evident, the rings should be replaced since the patches indicate the blow-by of gas. Check also that there is not a build-up of carbon behind the tapered ends of the rings, where they locate with the piston ring pegs.

8   It cannot be over-emphasised that the condition of the piston and rings in a two-stroke engine is of prime importance, especially since they control the opening and closing of the ports in the cylinder barrel by providing an effective seal. A two-stroke engine has only three working parts, one of which is the piston. It follows that the efficiency of the engine is very dependent on the condition of this component and the parts with which it is closely associated.

22 Small end - examination and renovation

The small end is a caged roller bearing and is checked at the same time as the crankshaft assembly.

23 Cylinder head - examination and renovation

1   It is unlikely that the cylinder head will require any special attention apart from removing the carbon deposit from the combustion chamber. Finish off with metal polish; a polished surface will reduce the tendency for carbon to adhere and will also help improve the gas flow.

2   Check that the cooling fins are not obstructed so that they receive the full air flow. A wire brush provides the best means of cleaning.

3   Check the condition of the thread where the spark plug is inserted. The thread in an aluminium alloy cylinder head is damaged very easily if the spark plug is overtightened. If necessary, the thread can be reclaimed by fitting what is known as a Helicoil insert. Most agents have facilities for this type of repair, which is not expensive.

4   If the cylinder head joint has shown signs of oil seepage when the machine was in use, check whether the cylinder head is distorted by laying it on a sheet of plate glass. Severe distortion will necessitate a replacement head but if the distortion is only slight it is permissible to wrap some emery cloth (fine grade) around the sheet of glass and rub down the joint using a rotary motion, until it is once again flat. The usual cause of distortion is uneven tightening of the cylinder head nuts.

24 Crankcases - examination and renovation

1   Inspect the crankcases for cracks or any other signs of damage. If a crack is found, specialist treatment will be required to effect a satisfactory repair.

2   Clean off the jointing faces using a rag soaked in methylated spirit to remove old gasket cement. Do not use a scraper because the jointing surfaces are damaged very easily. A leak-tight crankcase is an essential requirement of any two-stroke engine. Check also the bearing housings, to make sure they are not damaged. The entry to the housings should be free from burrs or lips.

25 Clutch actuating mechanism - examination

1   The clutch actuating mechanism and adjuster are attached to the generator cover. It is unlikely that these parts will require attention, particularly if the actuating mechanism has been greased regularly during routine maintenance.

2   Should it be necessary to dismantle the mechanism, unclip the return spring, unscrew the adjuster locknut and press the actuating lever out of its bush. The lever works on the quick-start worm principle and is unlikely to give trouble unless it is under-lubricated.

26 Clutch assembly - examination and renovation

1   Examine the condition of the linings of the fibre clutch plates. If they are damaged, loose or have worn thin, replacements will be required.

3   Examine the tongues of the plain clutch plates, where they engage with the clutch drum. After an extended period of service, burrs will form on the edges of the tongues which will correspond with grooves worn in the clutch drum slots. These burrs must be removed, by dressing with a smooth file.

3   The grooves worn in the clutch drum slots can be dressed in a similar manner, making sure that the edges of the slots are square once again. If this simple operation is overlooked, clutch troubles will persist because the plates tend to lodge in the grooves when the clutch is withdrawn and promote clutch drag.

4   Check the condition of the clutch springs. They should have a free length of $1^{1}/16$ inch and must be replaced if they compress much below this figure.

5   The clutch pushrod is in two sections, separated by a ball

bearing; a long section of approximately 5¾ inches and a short length of approximately 2 inches. Replace either section of the rod if the ends show a tendency to bell out. This is usually a sign of insufficient free play in the actuating mechanism, which has caused the ends to press together and generate heat, which in turn has destroyed the temper and caused them to soften and wear rapidly. The need for continuous clutch adjustment can invariably be traced to this type of fault.

27 Primary gear pinions - examination

Both primary gear pinions should be examined closely to ensure there is no damage to the helical teeth. The depth of mesh is predetermined by the bearing locations and cannot be adjusted.

28 Reassembly - general

1 Before the engine, clutch and gearbox components are reassembled, they must be cleaned thoroughly so that all traces of old oil, sludge, dirt and gaskets are removed. Wipe each part clean with a dry, lint-free rag to make sure that there is nothing to block the internal oilways of the engine.
2 Lay out all the spanners and other tools likely to be required so that they are close at hand during the reassembly sequence. Make sure the new gaskets and oil seals are available - there is nothing more infuriating than having to stop in the middle of a reassembly sequence because a gasket or some other vital component has been overlooked.
3 Make sure that the reassembly area is clean and unobstructed and that an oil can with clean engine oil is available so that the parts can be lubricated before they are reassembled. Refer back to the torque wrench settings and clearance data where necessary. Never guess or take a chance when this data is available.
4 Do not rush the reassembly operation or follow the instructions out of sequence. Above all, do not use excess force when parts will not fit together correctly. There is invariably good reason why they will not fit, often because the wrong method of assembly has been used.

29 Engine reassembly - fitting bearings to crankcase

1 Before fitting the crankcase bearings, make sure that the bearing surfaces are scrupulously clean and that there are no burrs or lips on the entry to the housings. Press or drive the bearings into the cases, using a mandrel and hammer, after first making sure that they are lined up squarely. Warming the crankcases will help when a bearing is a particularly tight fit.
2 When the bearings have been driven home, lightly oil them and make sure they revolve freely. This is particularly important in the case of the main bearings. There are two bearings on the drive side and one bearing on the generator side (right hand and left hand crankcases respectively) each of the journal ball type.
3 Using a soft mandrel, drive the oil seals into their respective locations. Do not use more force than is necessary because the seals will damage very easily. Good crankcase seals are essential to the efficient running of any two-stroke engine and if there is any doubt about the condition of the old seals they should be replaced without hesitation. Poor starting and indifferent running can often be attributed to worn or damaged oil seals, which allow air to enter the crankcase and dilute the incoming mixture whilst it is under crankcase compression.
4 Replace the bearing retaining plate on the gearbox main bearing.

30 Engine reassembly - fitting the crankshaft assembly

Feed the crankshaft assembly into the left hand crankcase

ensuring that the crankshaft shims are in place. Find suitable spacers and the rotor nut and draw the crankshaft assembly into the crankcase. Several spacers will be required as the thread length for the rotor nut is relatively short, so several 'bites of the cherry' will be required.

31 Engine reassembly - gearbox components

1 Support the crankcase on blocks.
2 Put the mainshaft and layshaft in mesh in one hand, rest the selector drum on top, pyramid fashion, feed the mainshaft selector fork into its groove and rotate it to put the roller into the cam track on the selector drum. Repeat for the layshaft selector. Hold everything in place with the other hand and feed the layshaft into its bearing. The mainshaft will find its bearing next, followed by the selector pins and the selector drum. The gear cluster should slide in, but if difficulty is experienced, check that the selector drum is in neutral and gently ease the components in. Do not use excessive force. Check that the shafts rotate.

32 Engine reassembly - rejoining the crankcases

1 Smear the joint face with Golden Hermatite or other jointing compound.
2 Check that the crankshaft shims are in position.
3 Lower the right hand crankcase onto the shafts. Gentle tapping may be required to push the crankshaft into its bearing. Note that the two hollow dowels must locate in the right hand crankcase.
4 When the crankcases are together, rotate the shafts to ensure that no binding occurs.
5 Excessive force should not be used as this shows that something has been wrongly assembled or is out of alignment.
6 Replace the twelve screws in their marked positions and refit the neutral light switch.
7 The layshaft spacer can then be pushed into the oil seal on the left-hand side of the engine. Use great care so that the seal is not damaged.
8 Replace the oil drain plug after checking that the sealing washer is in good condition.

33 Engine reassembly - reassembling the disc valve

1 Replace the drive pin in the crankshaft and slide on the drive collar.
2 Oil the surface on which the valve disc runs to lubricate it and to form a seal to reduce oil leaks.
3 Place the valve disc onto the drive collar ensuring that the two spots on the disc line up with the pin in the crankshaft.
4 A small amount of oil should be put on the disc for the same reasons as before.
5 The disc cover plate is then replaced, ensuring that the carburettor stub lines up with the hole in the crankcase, and the six retaining screws inserted and tightened.
6 Replace the spacer, helical gear and belleville washer (convex face outwards) and lock to the crankshaft with the nut.

34 Engine reassembly - reassembling the gearchange mechanism and index arm

1 Hook the index arm spring into the bearing retaining plate and the index arm and pull the index arm across to allow the shouldered bolt to screw into the crankcase.
2 Ensure that the gearchange return and pawl springs are in position. Feed the shaft through the crankcases, pulling the pawl clear of the selector drum whilst easing the gearchange spring over its pin to allow the gearchange assembly to seat home.
3 Replace the washer and circlip on the gearshaft on the other

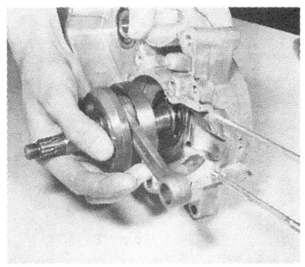

30.1a Feed crankshaft assembly into left-hand crankcase first

30.1b Use spacer assembly to draw crankshaft into position without damage

31.2a Mesh gear trains prior to assembly and fit with gear selector drum

31.2b Feed in mainshaft selector first, then layshaft selector before fitting whole assembly in crankcase

32.1 Check the gear shafts rotate before fitting other crankcase half

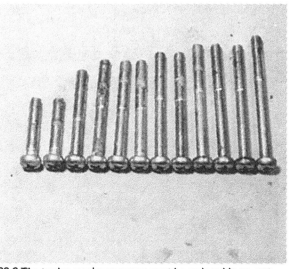

32.6 The twelve crankcase screws must be replaced in correct order

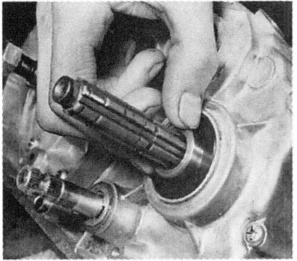
32.7 Push spacer through layshaft oil seal with great care

33.1a Replace drive pin in end of crankshaft and ...

33.1b ... slide on drive collar to engage with pin

33.2 Lubricate surface before fitting disc valve

33.3 Correct location of disc valve in relation to crankshaft is essential

33.5 Fit cover plate ensuring carburettor stub lines up with intake

33.6a Replace spacer, followed by ...

33.6b ... helical drive pinion, then fit nut and ...

33.6c ... tighten to recommended torque setting

34.1 Refit gear indexing arm

34.2 Feed gear change shaft through crankcases and engage pawl with selectors

34.3 Replace washer and circlip on end of gear change shaft

side of the engine.

4   Select second gear and check the relationship of the two claws of the gearchange shaft pawl and the uppermost pair of the five pins set in the end of the selector drum. The clearance between either claw and its respective pin must be exactly the same as the clearance between the other claw and pin, so that the gearchange assembly is correctly centralised. If adjustment is necessary, slacken its locknut and rotate carefully the adjusting screw set in the crankcase wall between the two straight ends of the shaft return spring; when adjustment is correct, tighten securely the locknut.

35 Engine reassembly - reassembling the kickstarter mechanism

1   Replace the thrust washer, sleeve and gear (with dogs uppermost) on the mainshaft.
2   Assemble on the layshaft a plain washer, the crinkle washer, the idler pinion and another plain washer. A circlip retains them all on the layshaft.
3   Reassemble the return spring sub-assembly by putting the small spring onto the shaft and put the gear on the shaft, ensuring that the spring sits in its correct position and retains the gear with the circlip. Assemble the plain spring register and the return spring ensuring that the end of the return spring is located in the hole in the shaft. The slotted spring register can now be fed into the spring and retained with the circlip.
4   The return spring sub-assembly can now be placed in the crankcase ensuring that the small spring locates in the special cast slot. The return spring is then hooked onto its pillar.

36 Engine reassembly - reassembling the clutch

1   The clutch drum is placed on the sleeve on the mainshaft. A slight anticlockwise twist is needed to mesh the helical gears and to lock the dogs on the clutch drum with the dogs on the gear.
2   The thrust plate and washer (where fitted), clutch centre, tab washer and nut are then assembled. Lock the clutch centre with a piece of metal and tighten the nut. Prise up the tab washer to lock the nut.
3   The plain portion of the pushrod is then pushed down the centre of the mainshaft, followed by the small ball bearing and the headed pushrod portion. The oil seal on the other side of the engine should hold the plain portion of the pushrod in position.
4   The clutch plates are assembled in this order: a fibre plate locating in the clutch drum, a plain aluminium plate locating on the clutch centre and another fibre plate.
5   The pressure plate is then assembled and the four springs and screws which hold everything in place. The screws should be tightened compressing the springs until the head contacts the pressure plate. If these screws are not tightened sufficiently, the clutch will slip when power from the engine is applied. If they are tightened unevenly the pressure plate will not disengage squarely and clutch drag will occur when the clutch lever is pulled in.
6   Replace the two O rings, one on the carburettor stub and one on the boss adjacent to it. Refit the right hand cover, making sure that the dowels fit correctly, and replace the seven screws noting that the long one is fitted adjacent to the carburettor stub.

37 Engine reassembly - replacing the final drive sprocket

1   Check that the spacer has been fitted and is properly seated in the oil seal. Refit the neutral indicator lamp switch and replace the plug which seals off the end of the gear selector drum.
2   Onto the layshaft slide the sprocket, tab washer and locking sleeve complete with the nut.
3   Locate the two half collets in the groove in the layshaft and finger-tighten the nut ensuring that the two half collets locate in the recess in the end of the sleeve.
4   Lock the engine and fully tighten the nut.
5   Prise the tab washer up to lock the nut.
6   Refit the circlip on the layshaft.

35.1a Thrust washer and sleeve fit over gearbox mainshaft, followed by ...

35.1b ... pinion with dogs uppermost

35.2a Plain washer and crinkle washer fit below idler pinion ...

35.2b ... plain washer and retaining circlip on top of pinion

35.3 Kickstarter return spring assembly fits next

35.4 Loop arm of return spring around extending pillar

36.1 Place clutch drum over sleeve so that dogs engage with those of pinion

36.2a Do not omit thrust plate before fitting clutch centre

36.2b Lock clutch centre so that nut can be fully tightened

36.3 Don't forget to insert the push rod assembly at this stage

36.4a Replace the clutch plates in their correct order ...

36.4b ... refit the clutch pressure plate and ...

36.4c ... tighten the springs evenly until their heads contact the pressure plate

36.6a Replace the two carburettor 'O' rings before ...

36.6b ... refitting the end cover

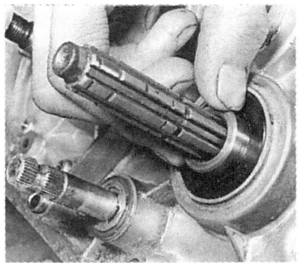

37.1a Check the spacer is seating correctly

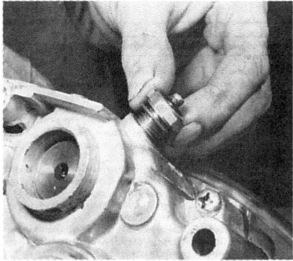

37.1b Refit the neutral indicator lamp switch

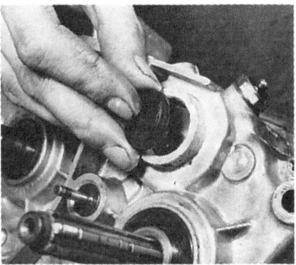

37.1c Do not omit the seal for the gear selector drum

37.3 Locate the two collets in the end of the threaded sleeve

37.4 Lock the engine and fully tighten the nut

38.1 Replace the Woodruff key in the end of the crankshaft

38 Engine reassembly - fitting the flywheel generator

1   The Woodruff key should now be tapped into the crankshaft.

2   The stator and its two fixing screws are then replaced, making sure that the wires feed out at the top of the engine so that the rubber grommet can be pushed back in its slot.

3   The neutral warning light switch can now be reconnected.

4   Before fitting the flywheel rotor, place a few drops of light oil on the felt wick which lubricates the contact breaker cam in the centre of the flywheel rotor.

5   It is advisable to check also whether the contact breaker points require attention at this stage, otherwise it will be necessary to withdraw the flywheel rotor again in order to gain access. Reference to Chapter 3 will show how the contact breaker points are renovated and adjusted.

6   Feed the rotor onto the crankshaft so that the slot lines up with the Woodruff key. The rotor may have to be turned to clear the heel of the contact breaker before it will slide fully home.

7   The washer and rotor nut can now be fitted and the nut fully tightened.

39 Engine reassembly - fitting the piston, cylinder barrel and cylinder head

1   Raise the connecting rod to its highest point and pad the mouth of the crankcase with clean rag as a precaution against displaced parts falling into the crankcase. Replace the caged roller small end bearing. Fit the piston, complete with rings, after oiling both the gudgeon pin and the small end bearing. To aid replacement in the correct position the piston crown is marked with an arrow head. This must face in the direction of the exhaust port or downward.

2   If the gudgeon pin is a tight fit in the piston bosses, warm the piston first. This will expand the metal and make the refitting easier.

3   Replace both circlips, making sure that they are fully engaged with their retaining grooves. A displaced circlip can cause severe engine damage. Never re-use the original circlips; always fit new replacements.

4   Fit a new cylinder base gasket. No gasket cement is required at this joint.

5   Smear the cylinder bore with clean oil and lower the cylinder barrel down the holding down studs until it can be engaged with the piston. Check that the piston rings are correctly lined up with the piston pegs and then compress the rings so that the piston can be inserted into the cylinder bore. The spigot of the cylinder barrel has a taper, to facilitate the entry of the piston rings. Do not use force; the piston rings are brittle and will break very easily if they become trapped.

6   When the piston and rings have passed into the cylinder bore, remove the crankcase padding and slide the cylinder barrel and piston downward until the spigot of the cylinder barrel enters the mouth of the crankcase and the flange seats correctly on the new cylinder base gasket.

7   Fit a new cylinder head gasket to the top flange of the cylinder barrel. This is a plain copper gasket.

8   Gasket cement is not required at the cylinder head joint.

9   Lower the cylinder head into position. Replace the four cylinder head nuts and washers and tighten them in a diagonal sequence, a little at a time, until the recommended torque wrench setting is achieved. Check that the piston is free from any tight spots when the crankshaft is turned before proceeding to the next sequence of operations.

40 Refitting the engine/gearbox unit in the frame

1   Follow in reverse the procedure given in Section 5 of this Chapter. Fit the spark plug as soon as the engine unit is lifted into the frame and the engine bolts are replaced, to prevent dirt and foreign matter from dropping into the engine.

38.2 Make sure stator wires feed out at top of engine unit

38.3 Reconnect neutral warning lamp switch

38.6 Refit rotor to line up with key in crankshaft

2 Use a new copper ring gasket for the exhaust pipe joint. A leaktight exhaust system is essential for the correct running of the engine.

3 Refit the kickstarter and gear change levers so that they are positioned at the correct angles. These are easier to determine when the engine is in the frame.

4 Remove the gearbox oil filler plug from the top of the right hand crankcase and refill the gearbox with SAE 10W/30 SE engine oil. The filler cap has an integral dipstick; oil should be added until it reaches the groove when the machine is standing on level ground. Do not overfill.

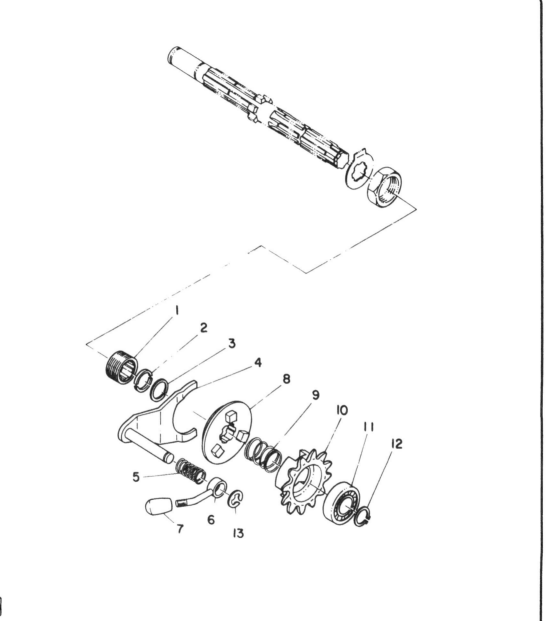

Fig. 1.9. Pedal engagement assembly

1	Threaded collar	4	Transmission change fork	7	Knob for transmission
2	Collets - 2 off	5	Lever return spring		change lever
3	Circlip	6	Transmission change lever	8	Transmission change dogs
				9	Transmission change spring

10	Drive sprocket from pedals (12 teeth)
11	Bearing
12	Circlip

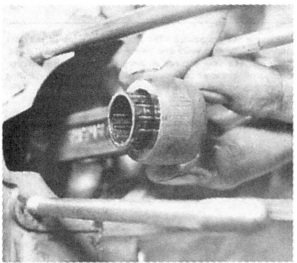

39.1a Replace the caged roller small end bearing

39.1b Refit the piston so that the arrow points to the exhaust port

39.3 Always fit new circlips - never re-use the originals

39.5 Smear cylinder barrel with oil before lowering into position

39.7 Use a new cylinder head gasket

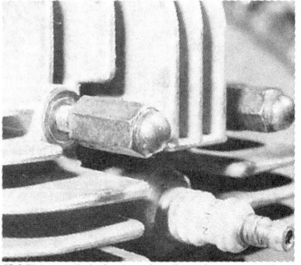

39.9 Tighten cylinder head nuts in a diagonal sequence to the recommended torque setting

40.1a Refit lower engine bolt first, to act as a pivot

40.1b Swing engine up into position and replace other engine bolts

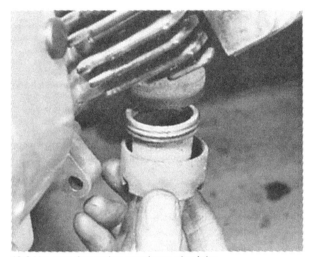

40.2 Use new ring gasket at exhaust pipe joint

41 Starting and running the rebuilt engine

1 When the initial start-up is made, run the engine slowly for the first few minutes, especially if the engine has been rebored. Check that all the controls function correctly and that there are no oil leaks before taking the machine on the road.

2 Remember that a good seal between the piston and the cylinder barrel is essential for the correct functioning of any two-stroke engine. In consequence, a rebored engine will require more careful running-in than its four-stroke counterpart. There is a far greater risk of engine seizure during the first hundred miles if the engine is permitted to work hard.

3 Do not add extra oil to the petrol/oil mix in the mistaken belief that it will aid running in. More oil means less petrol and the engine will run with a permanently weakened mixture, causing overheating and a far greater risk of engine seizure. Keep to the recommended proportions.

4 Do not tamper with the exhaust system or run the engine without the baffles fitted to the silencer. Unwarranted changes in the exhaust system will have a very noticeable effect on engine performance, invariably for the worst.

42 Fault diagnosis - engine

Symptom	Reason/s	Remedy
Engine will not start	Defective spark plug	Remove plug and lay on cylinder head. Check whether spark occurs when engine is kicked over.
	Dirty or closed contact breaker points	Check condition of points and whether gap is correct.
	Air leak at crankcase or worn oil seals around crankshaft	Flood carburettor and check whether mixture is reaching the spark plug.
	Clutch slip	Check and adjust clutch.
Engine runs unevenly	Ignition and/or fuel system fault	Check systems as though engine will not start.
	Blowing cylinder head gasket	Leak should be evident from oil leakage where gas escapes.
	Incorrect ignition timing	Check timing and reset if necessary.
	Loose pin on which moving contact breaker point pivots	Renew defective parts.

Lack of power	Incorrect ignition timing	See above.
	Fault in fuel system	Check system and filler cap vent.
	Blowing head gasket	See above.
	Choked silencer	Clean out baffles.
High fuel/oil consumption	Cylinder barrel in need of rebore and o/s piston	Fit new rings and piston after rebore.
	Oil leaks or air leaks from damaged gaskets or oil seals	Trace source of leak and replace damaged gaskets or seals.
Excessive mechanical noise	Worn cylinder barrel (piston slap)	Rebore and fit o/s piston.
	Worn small end bearing (rattle)	Renew bearing and gudgeon pin.
	Worn big-end bearing (knock)	Fit new big-end bearing.
	Worn main bearings (rumble)	Fit new journal bearings and seals.
Engine overheats and fades	Pre-ignition and/or weak mixture	Check carburettor settings. Check also whether plug grade correct.
	Lubrication failure	Is correct measure of oil mixed with petrol?

43 Fault diagnosis - clutch

Symptom	Reason/s	Remedy
Engine speed increases but machine does not respond	Clutch slip	Check clutch adjustment for pressure on pushrod. Also free play at handlebar lever. Check condition of clutch plate linings, also free length of clutch springs. Renew if necessary.
Difficulty in engaging gears. Gear changes jerky and machine creeps forward, even when clutch is fully withdrawn	Clutch drag	Check clutch adjustment for too much free play.
	Clutch plates worn and/or clutch drum	Check for burrs on clutch plate tongues or indentations in clutch drum slots. Dress with file.
	Clutch assembly loose on mainshaft	Check tightness of retaining nut. If loose, fit new tab washer and retighten.
Operation action stiff	Damaged, trapped or frayed control cable	Check cable and replace if necessary. Make sure cable is lubricated and has no sharp bends.
	Bent pushrod	Renew.

44 Fault diagnosis - gearbox

Symptom	Reason/s	Remedy
Difficulty in engaging gears	Pawl spring broken	Renew.
	Gear selector forks bent	Renew.
	Worn selector drum	Renew.
Machine jumps out of gear	Worn dogs on ends of gear pinions	Renew worn pinions.
	Index arm spring broken	Renew.
Kickstarter does not return when engine is turned over or started	Broken or badly tensioned kickstarter return spring	Renew.
Gear change lever does not return to normal position	Broken return spring	Renew.
Difficulty when changing gear	Broken index arm spring	Renew, using modified part

Chapter 2 Fuel system and lubrication

Refer to Chapter 7 for information relating to the 1975 on models

Contents

Specifications

Petrol tank capacity ... ... 6.0 litres (1.54/1.32 US/Imp gall)

Petrol/oil mixture:

Fuel grade	Unleaded, or leaded two-star (regular)
Oil grade	Good quality, self-mixing two-stroke oil
Mixing ratio	20:1, ie 0.4 (2/5) pt/227 cc oil to 1 gal petrol, 50 cc oil to 1 lit petrol

Carburettor:

Type	VM 16 SC
Main jet	150
Needle jet	W4
Needle	3G9 - 3
Pilot jet	25
Starter jet	50
Throttle slide cut away	1.5
Air screw setting	Back out 1¾ turns
Idling speed	1250 — 1350 rpm

1 General description

The fuel system comprises a petrol tank from which a petrol/oil mix of controlled proportions is fed by gravity to the float chamber of the carburettor. A petrol tap with a built-in gauze filter is located beneath the rear end of the petrol tank, which has provision for turning on a small reserve quantity of fuel when the main content of the tank is exhausted.

For cold starting purposes a handlebar control operates a movable slide to give a rich mixture.

2 Petrol/oil mix - correct ratio

1 Because the engine relies on the 'petroil' system for lubrication purposes, a measured amount of oil must always be added to the petrol. If a self-mixing oil is used, the proportion is one part of oil to twenty parts of petrol.
2 It will be realised that the lubrication of the engine is dependent solely on the intake of the fuel mixture from the carburettor. In consequence, it is inadvisable to coast the machine down a long hill whilst the throttle is closed, otherwise there is risk of engine seizure through the temporary lack of lubrication.
3 The gearbox has its own supply of lubricating oil and is quite independent of the engine lubricating system. Two-stroke mixing oils must NEVER be used.

3 Petrol tank - removal and replacement

1 It is unlikely that there will be need to remove the petrol tank completely unless the machine has been laid up and rust has formed inside or it needs reconditioning. The engine/gear unit can be removed from the frame without having to detach the tank. The ignition coil is mounted inside the frame and the tank has to be removed to gain access.
2 The petrol tank is supported by rubber buffers and is retained in place with the dualseat. There are two bolts at the front and two nuts at the rear fixing the dualseat to the machine. After the dualseat has been removed the rear of the petrol tank can be lifted and the tank slid off to the rear of the machine.

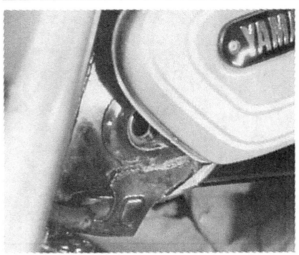

3.2a Rubbers on inside of tank engage with guides on frame

3.2b Lift rear end first when removing tank

3 To replace the tank reverse the procedure described in the preceding paragraph, after first ensuring that the front rubber buffers have slid into their correct positions.

4 Petrol tap - removal and replacement

1 The petrol tap is a three position closed - open - reserve type fitted to the left hand side of the petrol tank.
2 Drain the tank or if only a small amount of fuel remains, lay it on its right hand side. Undo the union nut above the tap and it will withdraw from the tank.
3 To clean the filter, the bolt on the bottom of the tap must be removed allowing the bottom half of the tap to fall off, revealing the filter to be cleaned. There is no necessity to remove or drain the tank when cleaning the filter, provided the petrol is turned off.
4 To replace the filter and cap reverse the procedure in paragraphs 2 and 3.

5 Petrol feed pipe - inspection

The petrol feed pipe is made of synthetic rubber and a check that it is not cracked or chafed should be made as leaking petrol can cause a fire. Ensure that the circlips on each end are present, in good condition and properly located.

6 Carburettor - removal

Engine removed from machine
1 If the engine has already been removed from the machine, the carburettor will be attached to the control cables and tied up out of the way.

Engine still in machine
2 If the engine is still in the machine, the carburettor is on the right hand side of the engine, under the front cover.
3 Turn off the petrol.
4 Remove the four screws and right hand cover to reveal the carburettor.
5 Slide the rubber boot and spring retainer up the control cables and pull the petrol pipe off.
6 Remove the plastic bung from the front of the cover, insert a screwdriver in the hole and slacken off the carburettor clamp ring. Pull the carburettor off its stub.

7 Carburettor - dismantling and inspection

1 At this stage, the carburettor is still attached to the control cables and to remove it completely the two small screws on the top of the carburettor must be undone. The top will come away complete with the throttle slide assembly and the choke slide.
2 To separate the float chamber, remove the screw from the bottom of the carburettor and gently pull off the float chamber, taking care not to damage the gasket or lose the O ring.
3 The carburettor float will stay in the float chamber when it is removed and can then be lifted or tipped out. Check the condition of the float and shake it to see if there is any petrol inside it. The float cannot be effectively repaired and if damaged, should be renewed. Make sure that the float chamber is clean and free from any sediment that may have originated from the petrol.
4 The float needle is situated inside the petrol pipe union and should be cleaned and checked for wear in the form of ridges in the conical portion of the end. Renew it if necessary.
5 The main jet is the smaller of the two brass hexagons on the underside of the carburettor body. It screws into the end of the needle jet and should be removed and cleaned by blowing or by screwing it into the end of a tyre pump and pumping it clear. Do not use pins or wire to clean it or the size and finish of the small hole will be affected.
6 The needle jet is the brass hexagon into which the main jet screws and should be removed for cleaning.
7 Clean the carburettor body and blow out the internal passages. Check for wear in the slide bores.
8 The throttle and air slide assemblies should be examined for wear. Check that the needle is straight and the return springs unbroken.
9 The air screw is the slotted screw on the front of the carburettor and can be unscrewed for cleaning.
10 When reassembling the carburettor, follow the dismantling instructions in reverse, ensuring that the small air pipe is refitted on its union on the front of the carburettor and the petrol overflow pipe on the side.

8 Carburettor - checking the settings

1 The various sizes of the jets and the throttle slide, needle and needle jet are predetermined by the manufacturer and should not require modification. Check with the Specifications list if there is any doubt about the values fitted.
2 Slow running is controlled by a combination of the throttle stop and air screw. Commence by screwing in the air screw until it lightly seats then back it out 1¾ complete turns. Next, adjust the throttle stop so that the engine idles at a steady tick-over.
3 As a rough guide, up to 1/8 throttle is controlled by the pilot jet, from 1/8 to ¼ throttle by the throttle slide cutaway, from ¼ to ¾ throttle by the needle position and from ¾ to full

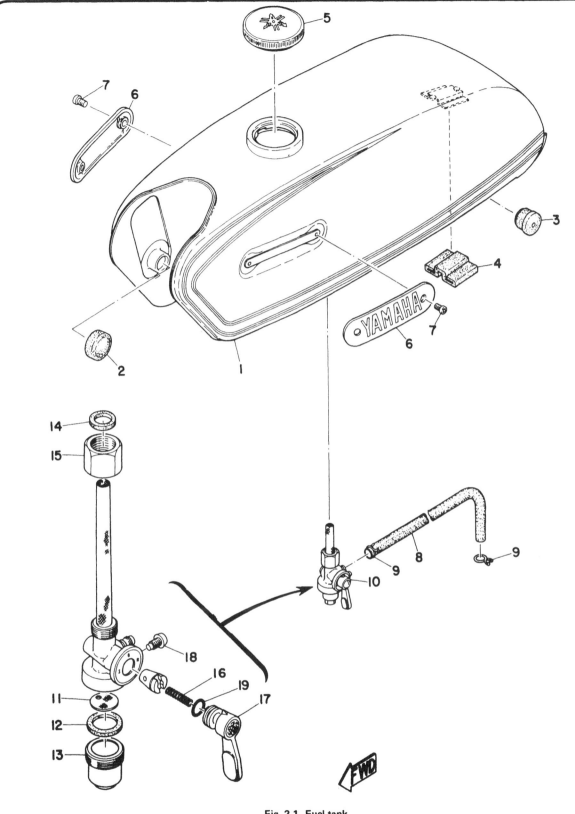

Fig. 2.1. Fuel tank

1 Fuel tank complete	6 Tank badges - 2 off	10 Fuel tap complete	15 Union nut
2 Locating rubber - 2 off	7 Panhead screw for tank	11 Filter gauze	16 Tap lever spring
3 Locating rubber - 2 off	badge - 4 off	12 Sealing washer	17 Tap lever
4 Locating rubber	8 Fuel pipe	13 Filter bowl	18 Tap lever fitting screw
5 Filler cap	9 Pipe clip - 2 off	14 Sealing washer	19 Seal for tap lever

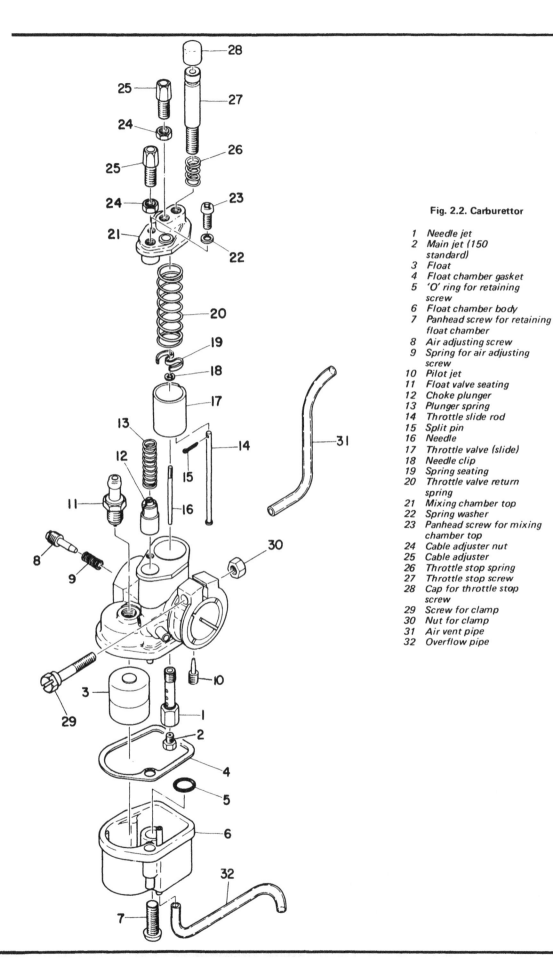

Fig. 2.2. Carburettor

1 Needle jet
2 Main jet (150 standard)
3 Float
4 Float chamber gasket
5 'O' ring for retaining screw
6 Float chamber body
7 Panhead screw for retaining float chamber
8 Air adjusting screw
9 Spring for air adjusting screw
10 Pilot jet
11 Float valve seating
12 Choke plunger
13 Plunger spring
14 Throttle slide rod
15 Split pin
16 Needle
17 Throttle valve (slide)
18 Needle clip
19 Spring seating
20 Throttle valve return spring
21 Mixing chamber top
22 Spring washer
23 Panhead screw for mixing chamber top
24 Cable adjuster nut
25 Cable adjuster
26 Throttle stop spring
27 Throttle stop screw
28 Cap for throttle stop screw
29 Screw for clamp
30 Nut for clamp
31 Air vent pipe
32 Overflow pipe

7.1 Throttle and air slide lift away with carburettor top

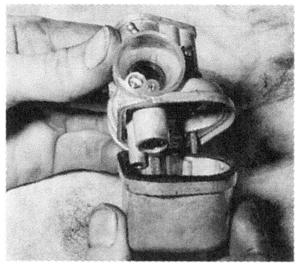

7.2 Release screw at base of carburettor to free float chamber

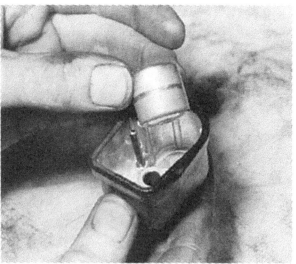

7.3 Float will remain in float chamber and will lift out

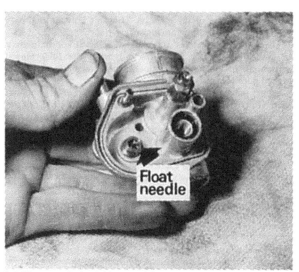

7.4 Float needle is within petrol feed union

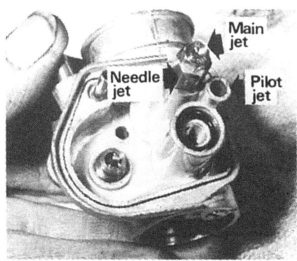

7.5 Main jet screws into base of needle jet

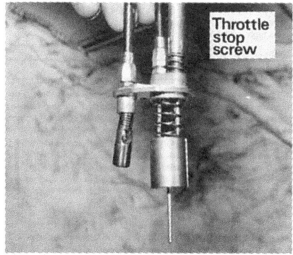

7.8 Examine throttle and air slide assembly for wear

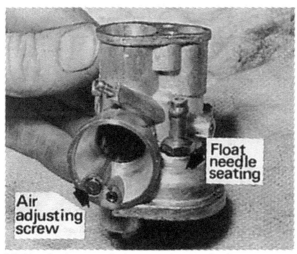

7.9 Air adjusting screw is close to carburettor intake

throttle by the size of the main jet. These are only approximate divisions; there is a certain amount of overlap.

4   The normal setting for the pilot jet screw is approximately one and three-quarter full turns out from the fully closed position. If the engine 'dies' at low throttle openings, suspect a blocked pilot jet.

5   Guard against the possibility of incorrect carburettor adjustments which result in a weak mixture. Two-stroke engines are very susceptible to this type of fault, which will cause rapid overheating and subsequent engine seizure. Some owners believe that the addition of a little extra oil to the petrol will help prolong the life of the engine, whereas in practice quite the opposite occurs. Because there is more oil the petrol content is less and the engine runs with a permanently weakened mixture!

9 Air cleaner - location, examination and replacement of the element

1   The air cleaner is above the engine and to remove the filter element for cleaning the nut in the centre of the chrome endplate must be removed. The two endplates and centre rod can then be removed. The filter element will then pull clear of the air hose and will slide out.

2   To clean the filter element blow it with an air line from the inside. As it is only made of paper it should be kept clear of water or oil. If the filter is torn or wet it should be renewed without question. A blocked or partially blocked air cleaner causes high petrol consumption.

10 Exhaust system - cleaning

1   The exhaust system of any two-stroke engine requires quite frequent attention because the oily nature of the exhaust gases causes a build-up of sludge which will eventually partially block the system and cause serious back pressures. This will occur even more rapidly if the engine is in need of a rebore and is using oil.

2   The exhaust system is removed easily by following the procedure detailed in Chapter 1.5, paragraph 17. The baffle tube is removed by unscrewing the retaining screw and sliding out the baffle tube. Always remove the wadding wrapped around the baffle tube.

3   The exhaust pipe and silencer are one unit and if a large amount of carbon has built up inside it is necessary to fill the silencer with a solution of caustic soda after blocking up one end. If possible, leave the caustic soda solution within the silencer overnight before draining off and washing out thoroughly with water.

4   Caustic soda is highly corrosive and every care should be

taken when mixing and handling the solution. Keep the solution away from the skin and more particularly the eyes. The wearing of rubber gloves is advised whilst the solution is being mixed and used.

5   The solution is prepared by adding 3 lbs of caustic soda to 1 gallon of COLD water, whilst stirring. Add the caustic soda a little at a time and NEVER add the water to the chemical. The solution will become hot during the mixing process, which is why cold water must be used.

6   Make sure the used caustic soda solution is disposed of safely, preferably by diluting with a large amount of water. Do not allow the solution to come into contact with aluminium castings because it will react violently with this metal.

7   If the baffle assembly is heavily coated with a sludge of carbon and oil, it is permissible to burn this out with a blow lamp. The wadding wrapped around the tube merely accelerates the build-up of carbon and should be removed as soon as possible.

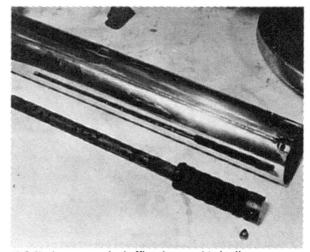

10.2 Single screw retains baffle tube assembly in silencer

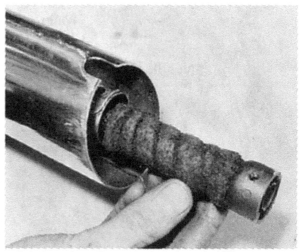

10.7 Wadding should be removed as soon as possible

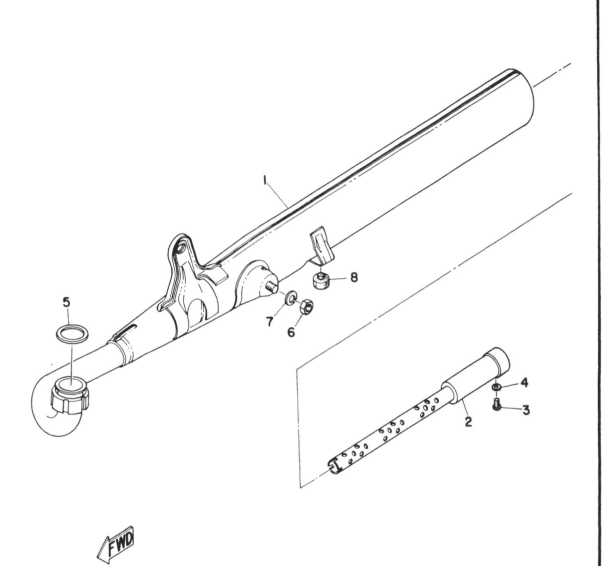

Fig. 2.3. Exhaust pipe and silencer

1 Exhaust assembly
2 Baffle tube

3 Panhead screw for retaining
  baffle assembly

4 Spring washer
5 Exhaust pipe gasket

6 Silencer mounting nut
7 Spring washer
8 Stop for centre stand

8   Never tamper with the exhaust system and remove the baffles from the silencer or fit a quite different system. Although a louder exhaust note may give the illusion of greater speed, in nearly every case performance will be reduced and the rider may risk prosecution. It is difficult to improve on the manufacturer's original specification, which has been designed to match in with the characteristics of the engine. Speed and noise do not necessarily go hand in hand.

9   To reassemble the exhaust system reverse the dismantling procedure.

11 Fault diagnosis - fuel system and lubrication

Symptom	Reason/s	Remedy
Excessive fuel consumption	Air cleaner choked or restricted	Clean or if paper element oily or wet renew.
	Fuel leaking from carburettor. Float sticking	Check all unions and gaskets. Float needle seat needs cleaning.
	Badly worn or distorted carburettor	Renew.
	Carburettor incorrectly adjusted	Tune and adjust as necessary.
	Incorrect silencer fitted to exhaust system	Do not deviate from manufacturer's original silencer design.
Idling speed too high	Throttle stop screw in too far. Carburettor top loose	Adjust screw. Tighten top.
Engine does not respond to throttle	Back pressure in silencer. Float displaced or punctured	Check baffles in silencer. Check whether float is correctly located or has petrol inside.
	Use of incorrect silencer or baffles missing	See above. Do not run without baffles.
Engine dies after running for a short while	Blocked air hole in filler cap	Clean.
	Dirt or water in carburettor	Remove and clean out.
General lack of performance	Weak mixture; float needle stuck in seat	Remove float chamber or float and clean.
	Air leak at carburettor joint or in crankcase	Check joints to eliminate leakage.
Excessive white smoke from exhaust	Too much oil in petrol, or oil has separated out	Mix in recommended ratio only. Mix thoroughly if mixing pump not available.

Chapter 3 Ignition system

Refer to Chapter 7 for information relating to the 1975 on models

Contents

Specifications

Flywheel generator:

Make	Mitsubishi or Hitachi
Type	FAZ - IQL or FII - L40
Output ignition winding	150 – 300 V
Contact breaker gap	0.012 - 0.016 in (0.30 - 0.40 mm)
Capacitor rating	0.22 uf ± 10%

Ignition coil:

Make	Mitsubishi
Type	HP - BI
Minimum spark gap	0.24 in (6 mm) @ 500 rpm

Spark plug: ... ... NGK B - 7HS

Spark plug gap	0.020 in. – 0.024 in. (0.5 – 0.6 mm)
Ignition timing fixed at	0.071 in. ± 0.006 in. (1.8 ± 0.15 mm) btdc

1 General description

The spark which is necessary to ignite the petrol/air mixture in the combustion chamber is derived from an ignition coil mounted on the frame and a generator attached to the left hand crankshaft of the engine. A contact breaker assembly within the generator determines the exact moment at which the spark will occur; as the points separate the electrical circuit is interrupted and a high tension voltage is developed across the points of the spark plug which jumps the air gap and ignites the mixture.

2 Flywheel generator - checking output

The output from the flywheel generator can be checked only with specialised test equipment of the multi-meter type. It is unlikely that the average owner/rider will have access to this equipment or instruction in its use. In consequence, if the performance of the generator is suspect, it should be checked by a Yamaha agent or an auto-electrical expert.

3 Ignition coil - checking, removal and replacement

1 The ignition coil is a sealed unit, designed to give long service. If a weak spark and difficult starting cause its performance to be suspect, it should be tested by an auto-electrical expert. A faulty coil must be replaced; it is not practical to effect a repair.

2 To remove the coil the petrol tank must first be removed as described in Chapter 2.3. The coil is mounted inside the frame and is retained with two nuts. These must be removed and the coil pulled out carefully by the high tension lead. The coil holder is then removed by undoing the two retaining screws.

3 Reassembly is the reverse of the removal procedure.

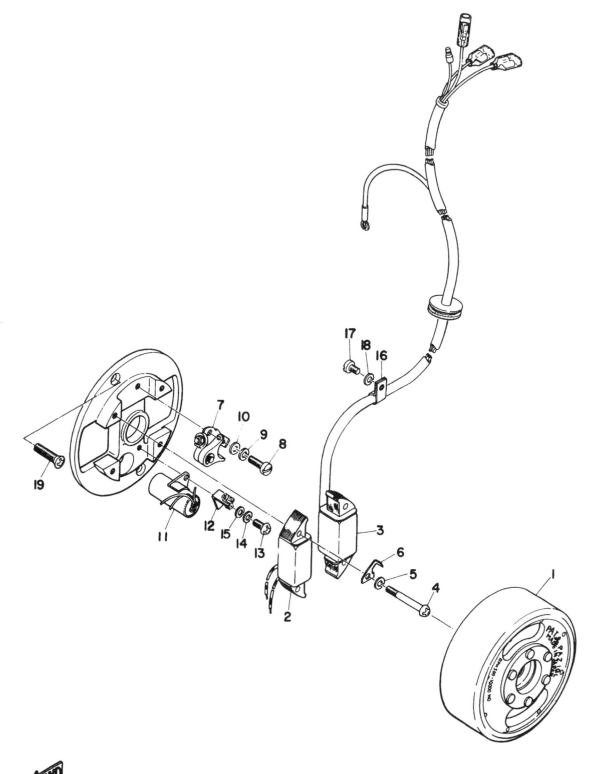

Fig. 3.1. Flywheel generator - Mitsubishi type (Hitachi similar)

1 Rotor	5 Spring washer - 4 off	10 Plain washer	15 Plain washer
2 Coil 1	6 Timing plate	11 Condenser	16 Cable clamp
3 Coil 2 (lighting)	7 Contact breaker assembly	12 Lubricating wick	17 Panhead screw for clamp
4 Panhead screw for retaining coil - 4 off	8 Panhead screw	13 Panhead screw	18 Spring washer
	9 Spring washer	14 Spring washer	19 Flathead screw for stator plate - 2 off

4 Contact breaker - adjustment

1 To gain access to the contact breaker assembly, remove the two screws which hold the flywheel generator cover in position and lift off the cover. The contact breaker points can be viewed through one of the apertures in the flywheel rotor.

2 Rotate the engine until the contact breaker points are in the fully open position. Examine the faces of the contacts. If they are pitted or burnt it will be necessary to remove them for further attention, as described in Section 5 of this Chapter.

3 The correct contact breaker gap, when the points are fully open, is 0.014 in (0.35 mm). Adjustment is effected by slackening the screw which clamps the fixed contact point in position and moving the contact nearer or further away, as the case may be, by levering against the timing plate. Make sure that the points are open fully when this adjustment is made, otherwise a false reading will result. Tighten the screw and check again.

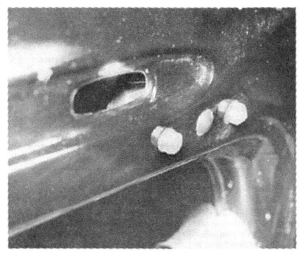

3.2 Ignition coil is housed within frame and retained by two nuts

5 Contact breaker points - removal, renovation and replacement

1 If the contact breaker points are burned, pitted or badly worn, they should be removed for dressing. If it is necessary to remove a substantial amount of material before the faces can be restored, new replacements should be fitted.

2 It is necessary first to withdraw the flywheel magneto rotor before access can be gained. Instructions for the removal of the rotor are given in Chapter 1, Section 7. The fixed contact is removed by withdrawing the screw which holds the assembly to the stator plate of the generator. The moving contact is detached by releasing the circlip from the end of the pivot pin and by freeing the leaf return spring from its point of attachment close to the lower coil.

3 The points should be dressed with an oilstone or fine emery cloth. Keep them absolutely square during the dressing operation, otherwise they will make angular contact when they are replaced and will burn away rapidly as a result.

4 Replace the contacts by reversing the dismantling procedure. Take particular care to replace any insulating washers in their correct sequence, otherwise the points will be isolated electrically and the ignition system will not function. Lightly grease the pivot pin before the moving contact is replaced and check that there is no oil or grease on the surfaces of the points.

5 Replace the flywheel rotor after greasing the internal contact breaker cam. It is also advisable to add a few drops of light oil to the lubricating wick which rubs on the contact breaker cam, if the wick has a dry appearance.

6 Re-adjust the contact breaker gap after the flywheel rotor has been locked in position and the centre retaining bolt tightened fully to the recommended torque wrench setting of 5.0 - 7.0 kgf m (36.0 - 50.5 lbf ft).

4.1 Access to points is through aperture in rotor

6 Condenser - removal and replacement

1 A condenser is included in the contact breaker circuitry to prevent arcing across the contact breaker points as they separate. It is connected in parallel with the points and if a fault develops, ignition failure will occur.

2 If the engine is difficult to start, or if misfiring occurs, it is possible that the condenser is at fault. To check whether the condenser has failed, remove the flywheel magneto cover and observe the points whilst the engine is running. If excessive sparking occurs across the points and they have a blackened or burnt appearance, it may be assumed the condenser is no longer serviceable.

3 The condenser is attached to the stator plate and it is first necessary to withdraw the flywheel rotor as described in Chapter 1, Section 7.

6.3 Condenser is attached to the stator plate

4   Before the condenser can be removed from the stator plate, it is necessary to unsolder the contact breaker and ignition coil leads. The retaining screw can then be removed and the condenser and wick holder will pull clear.

5   Fit the new condenser and the wick holder and retain them with the screw. Resolder the leads onto the new condenser.

6   Replace the flywheel rotor and lock it in position before tightening the centre retaining nut. Complete the reassembly by fitting the flywheel cover.

7   Ignition timing: checking and resetting

1   Certain items of specialised equipment are needed to check the ignition timing, these being rather expensive and infrequently required. It is therefore recommended that the machine be taken to a Yamaha Service Agent for the ignition timing to be checked. The items required are: a dial gauge set (Yamaha Part Number 90890-01173), comprising an accurate dial gauge, a suitable extension rod with a ball-point tip and an adaptor suitable for a 14 mm spark plug thread, and a self-powered points checker (Yamaha Part Number 90890-03031). This last item can be replaced by a proprietary multimeter set to the x 1 ohm resistance scale or by a battery and bulb test circuit. For those who have the necessary equipment, proceed as follows.

2   Remove the spark plug and the flywheel generator cover. Remove the left-hand sidepanel and disconnect the black or black/white wire which leads from the points to the ignition switch and the H.T. coil. Fit the dial gauge adaptor in the spark plug thread and tighten it securely, then screw the extension rod into the gauge, and insert the gauge assembly into the adaptor, securing it with the grub screw. Rotate the flywheel until the piston is at top dead centre (TDC); as the piston approaches TDC the gauge reading will decrease, stop momentarily as TDC is reached, then increase again as the piston descends. Set the gauge to zero as TDC is reached, then rock the flywheel to and fro to make sure that the needle does not go past zero.

3   The exact time at which the contact breaker points open is determined by the use of a points checker or a multimeter; the gauge needle will swing from 'Closed' to 'Open' (points checker) or will flicker to indicate increased resistance (multimeter). A battery and bulb test circuit can be used so that the bulb lights when the points are closed; note, however, that the bulb will not go out, but will merely glow dimmer as the points open. To make this more obvious to the eye, a high-wattage bulb must be used.

4   Connect the meter positive (+) terminal to the wire leading from the points and the meter negative (−) terminal to a good earth point on the engine. If a battery and bulb are used, obtain three lengths of wire, connect the battery negative (−) terminal to a good earth point, the battery positive (+) terminal to the bulb contact and the bulb body to the wire leading from the points.

5   Rotate the flywheel clockwise until a reading of 0.08 - 0.12 in (2-3 mm) is shown on the gauge, then rotate it slowly anti-clockwise until the points open. The reading shown on the gauge should be precisely 0.071 in (1.8 mm) although a tolerance of 0.006 in (0.15 mm) is allowed on either side of the set figure and any reading within these limits is permissible.

6   The setting is adjusted by opening or closing the contact breaker points gap to advance or retard respectively the ignition timing. Repeat the procedure given in paragraph 5 to check that the timing is now correct.

7   When the timing is found to be correct, measure very carefully the points gap. If it is found to be outside the permitted tolerance of 0.012 - 0.016 in (0.30 - 0.40 mm) the contact breaker points are excessively worn, either on the contact faces or on the heel of the moving contact, and must be renewed. It is essential that both the contact breaker points gap and the ignition timing setting are kept within the limits specified for each.

8   Working as described in Section 5 of this Chapter, fit a new set of contact breaker points; note that it is essential that only genuine Yamaha points should be used. Refit the flywheel and set the points gap to exactly 0.014 in (0.35 mm), then repeat the procedure given in paragraph 5 above. The ignition timing should be correct, or at least within tolerances.

9   Note that in some cases, depending on model, year and the type of flywheel generator fitted, marks are provided which apparently make it possible to check the ignition timing using a timing lamp or strobe. These marks may take the form of a line stamped on the outside of the rotor aligning with a mark cast or stamped on the crankcase cover or wall, or a fixed pointer on the stator plate (usually soldered to the condenser) which aligns with marks stamped on the rim of the rotor apertures. If such marks are provided, their accuracy must be checked with a dial gauge, as described above, before a timing lamp is used to check the ignition timing dynamically (ie with the engine running). Note that the machine's manufacturer only recommends the static method as being sufficiently accurate.

8   Spark plug - checking and resetting the gap

1   An NGK B-7HS plug is fitted as standard equipment to the Yamaha.

2   The spark plug has a 14 mm thread. The recommended gap is 0.020 - 0.024 inch (0.5 - 0.6 mm). Always use the grade of plug recommended or the exact equivalent in another manufacturer's range.

3   Check the spark plugs gap every six months or 1000 miles, whichever is soonest. To reset the gap, bend the outer electrode to bring it closer to the central electrode, otherwise the insulator will crack, causing engine damage if particles fall in whilst the engine is running.

4   Always carry a spare spark plug in the tool kit, wrapped so that the electrodes cannot become bent or dirt enter the electrode area. Two-stroke engines are more susceptible to plug troubles and may on occasion cause plug fouling or whiskering. A spare plug will enable the engine to be restarted promptly in the event of trouble, which invariably occurs at an ill-timed moment.

5   The condition of the spark plug electrodes and insulator can be used as a reliable guide to engine operating conditions. See accompanying diagrams.

6   Never overtighten a spark plug, otherwise there is danger of stripping the threads from the cylinder head, particularly those cast in light alloy. The plug should be sufficiently tight to seat firmly on copper sealing washer. Use a spanner which is a good fit, otherwise the spanner may slip and break the ceramic insulator.

7   If the thread within the cylinder head does strip, it can be repaired permanently and economically by the use of a Helicoil thread insert. Many dealers have facilities for effecting this type of repair at a fraction of the cost of a replacement cylinder head.

8   Make sure that the plug insulating cap is a good fit. This cap contains the suppressor which eliminates radio and TV interference.

Note: fault diagnosis appears on page 60

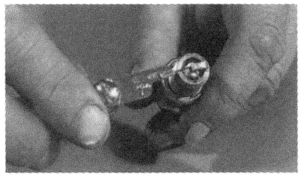

Spark plug maintenance: Checking plug gap with feeler gauges

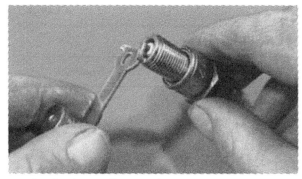

Altering the plug gap. Note use of correct tool

Spark plug conditions: A brown, tan or grey firing end is indicative of correct engine running conditions and the selection of the appropriate heat rating plug

White deposits have accumulated from excessive amounts of oil in the combustion chamber or through the use of low quality oil. Remove deposits or a hot spot may form

Black sooty deposits indicate an over-rich fuel/air mixture, or a malfunctioning ignition system. If no improvement is obtained, try one grade hotter plug

Wet, oily carbon deposits form an electrical leakage path along the insulator nose, resulting in a misfire. The cause may be a badly worn engine or a malfunctioning ignition system

A blistered white insulator or melted electrode indicates over-advanced ignition timing or a malfunctioning cooling system. If correction does not prove effective, try a colder grade plug

A worn spark plug not only wastes fuel but also overloads the whole ignition system because the increased gap requires higher voltage to initiate the spark. This condition can also affect air pollution

9 Fault diagnosis - ignition system

Symptom	Reason/s	Remedy
Engine will not start	No spark at plug	Try replacement plug if gap correct.
		Check whether contact breaker points are opening and closing, also whether they are clean.
		Check whether points arc when separated. If so, replace condenser.
		Check ignition switch and coil.
Engine starts but runs erratically	Intermittent or weak spark	Try replacement plug. Check whether points are arcing. If so, replace condenser.
		Check accuracy of ignition timing.
		Low output from flywheel magneto generator or imminent breakdown of ignition coil.
		Plug has whiskered. Fit replacement.
		Plug lead insulation breaking down. Check for breaks in outer covering, particularly near frame.

Chapter 4 Frame and forks

Refer to Chapter 7 for information relating to the 1975 on models

Contents

1 General description

The frame is of the spine type with swinging arm rear suspension controlled by two hydraulically damped suspension units. The front fork assembly comprises spring loaded fork tubes of the telescopic variety.

Repairs to the frame are limited to replacement of the parts which are subject to wear. The frame is constructed from pressed sheet metal and it is not economically feasible to repair accident damage in view of the low cost of a replacement.

2 Front forks - removal from frame

1 It is extremely unlikely that the front forks will need to be removed from the frame as a unit unless the steering head bearings give trouble or the forks are damaged in an accident.
2 Commence operations by placing the machine on the centre stand.
3 Slacken the front brake cable and disconnect it from the brake. Open the cable clip on the bottom yoke and remove the cable. Disconnect the cable from the handlebars and hang it up for the routine oiling procedure.
4 Remove the circlip retaining the speedometer cable in the brake plate and pull the cable clear. Unhook the cable from the clip and unscrew it from the speedometer head.
5 Undo the front wheel spindle nut and withdraw the wheel spindle. Support the machine to stop it from toppling forward.
6 The front wheel will now drop from the forks. Remove the spacer on the left hand side of the wheel to avoid it being lost.
7 Remove the bolts and washers from the top of the fork legs which will allow the speedometer and its bracket to be removed.
8 Slacken the screw on the bottom of the headlamp and prise off the headlamp rim.
9 Disconnect the wires to the headlamp bulb, the horn and the front indicators and place the reflector unit in a safe place.
10 Unscrew the nuts inside the headlamp shell and remove the

indicators. The headlamp shell will then rest on the bottom yoke.
11 The handlebars can be lifted clear once the bolts and half clamps have been removed.
12 Undo the bolt in the centre of the steering head stem and remove the washer and top yoke.
13 A large slotted nut is revealed and before this is unscrewed provision should be made to catch the uncaged ball bearings. There is a total of 38, 19 in each race.
14 Unscrew the large slotted nut whilst supporting the forks in position, remove the dust cover and the cone of the top race. The ball bearing can then be removed either with a magnet or a greased screwdriver.
15 As the forks are lowered the balls in the lower race will be displaced and once these have been collected the fork assembly can be pulled clear of the frame.
16 If further dismantling is necessary the front mudguard can be removed, when the forks have been withdrawn, by undoing the four retaining bolts.
17 The headlamp brackets can also be slid off the fork legs.

3 Front forks - dismantling

1 If only the fork legs are to be removed without disturbing the head races, follow the instructions in the previous heading as far as paragraph 7, and include paragraph 6. Then continue as follows:
2 To remove the fork legs, unscrew the pinch bolts in the fork bottom yoke. The fork legs should now pull clear. If they are still a tight fit, spring open the pinch bolt joint a little.
3 The O ring in the top of the fork leg should be removed to avoid its being lost and the oil drained out of the leg.
4 The rubber gaiter can now be slid off the fork leg with the spring inside it. The spring can easily be pulled out of the gaiter for checking or replacing.
5 The spring guide and register can be slid off the fork leg.
6 Unscrew the screwed collar and remove it. This will allow

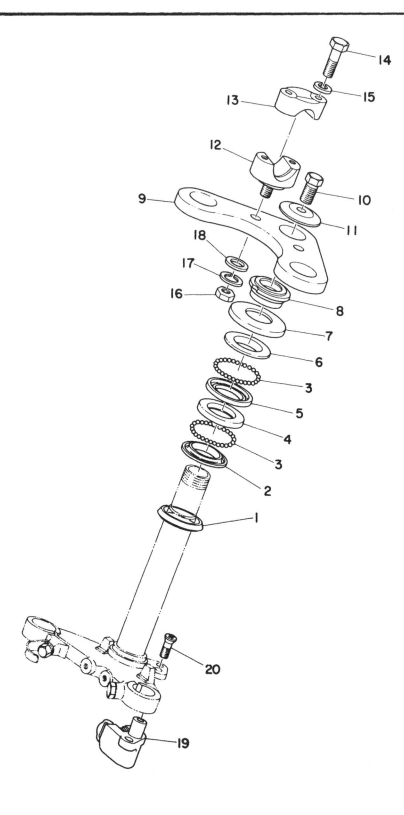

Fig. 4.1. Steering head assembly

1	Dust seal	6	Upper cone	12	Lower handlebar clamp - 2 off	16	Clamp mounting nut - 2 off
2	Lower cone	7	Dust cover	13	Upper handlebar clamp - 2 off	17	Spring washer - 2 off
3	Ball bearing (¼ inch) - 38 off	8	Adjusting nut	14	Clamp bolt - 4 off	18	Plain washer - 2 off
4	Lower cup	9	Top fork yoke	10	Top fork bolt - 2 off	15	Spring washer - 4 off	19	Steering column lock
5	Upper cup	11	Fork bolt washer - 2 off			20	Ovalhead screw for lock

1 Dust seal
2 Lower cone
3 Ball bearing (¼ inch) - 38 off
4 Lower cup
5 Upper cup

6 Upper cone
7 Dust cover
8 Adjusting nut
9 Top fork yoke
10 Top fork bolt - 2 off
11 Fork bolt washer - 2 off

12 Lower handlebar clamp - 2 off
13 Upper handlebar clamp - 2 off
14 Clamp bolt - 4 off
15 Spring washer - 4 off

16 Clamp mounting nut - 2 off
17 Spring washer - 2 off
18 Plain washer - 2 off
19 Steering column lock
20 Ovalhead screw for lock

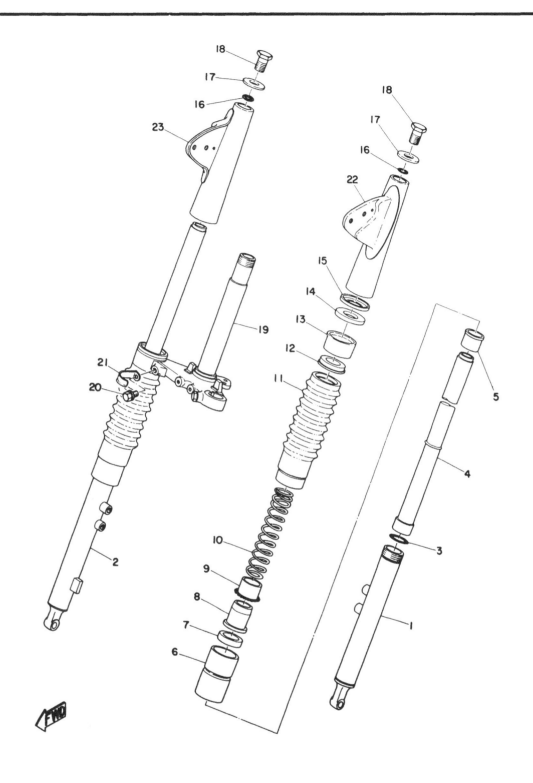

Fig. 4.2. Front forks

1	Lower fork leg - left-hand	6	Screwed collar - 2 off
2	Lower fork leg - right-hand	7	Oil seal - 2 off
3	'O' ring - 2 off	8	Fork spring seating - 2 off
4	Stanchion - 2 off	9	Spring guide - 2 off
5	Upper fork bush - 2 off	10	Fork spring - 2 off
		11	Gaiter - 2 off

12	Fork spring seating - upper - 2 off
13	Outer cover - 2 off
14	Packing piece - 2 off
15	Top cover guide - 2 off
16	'O' ring seal - 2 off
17	Fork cap washer - 2 off

18	Fork cap - 2 off
19	Head stem complete
20	Lower yoke pinch bolt - 2 off
21	Cable clamp
22	Top cover - left-hand
23	Top cover - right-hand

the inner stanchion to be pulled out of the fork leg which will pull the bush out of the fork leg. The bush can then be slid off the stanchion.

7   An O ring is fitted on the fork leg below the thread which can now be removed.

4 Front forks - general examination

1   Apart from the oil seals and bushes, it is unlikely that the forks will require any additional attention unless the fork springs are weak or if the fork legs or yokes have been damaged in an accident.

2   Visual examination will show whether the fork yokes are distorted or if the inner fork tubes are bent. It is rarely possible to effect a satisfactory repair and replacement is strongly recommended.

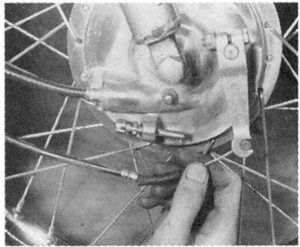

2.3 Disconnect cable from front brake operating arm

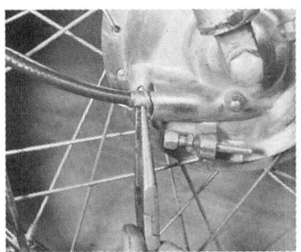

2.4a Remove circlip retaining speedometer drive cable and ...

2.4b ... pull cable from housing to disconnect drive

2.5 Front wheel spindle will pull out

2.6 Front wheel is now free to be taken away

2.7 Remove bolts and washers from top of each fork leg

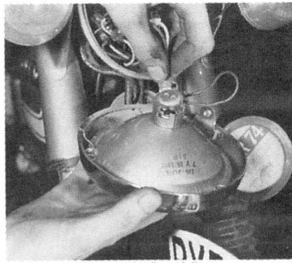

2.9 Disconnect wires within headlamp shell

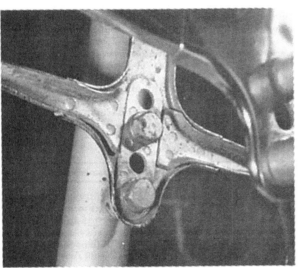

2.16 Front mudguard is bolted to inside of each fork leg

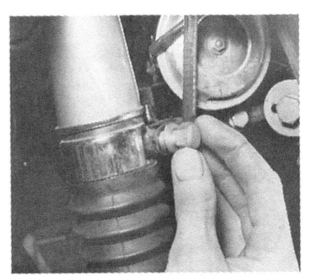

3.2 Remove pinch bolt from lower fork yoke

3.3 Remove 'O' ring from top of fork leg

3.4 Slide off rubber gaiter with spring within

3.5 Fork spring guide and register will slide off fork leg

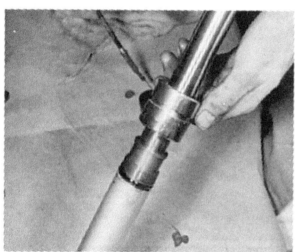

3.6a Unscrew plated collar and then ...

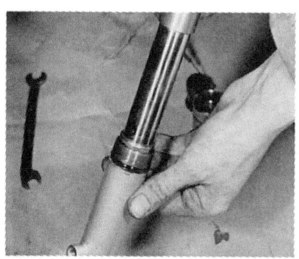

3.6b ... pull stanchion from lower fork leg

5 Front forks - examination and replacement of oil seals

1 If the fork legs have shown a tendency to leak oil or if there is any other reason to suspect the condition of the oil seals, now is the time to replace them.

2 The oil seals fitted to the plated cups are displaced quite easily by pushing them out of the cup. Note that the seals will be destroyed during removal, so new replacements are essential.

3 It is advisable to replace the O rings at the same time.

4 When fitting the replacement seals, coat the underside with jointing compound and enter each seal squarely into the holder with the 'open' side facing downward (closest to front wheel spindle). Complete the operation whilst the jointing compound is still wet.

6 Front forks - examination and replacement of bushes

1 Some indication of the extent of wear of the fork bushes can be gained before the machine is dismantled. If the front wheel is gripped between the knees and the handlebars rocked to and fro, the amount of wear will be magnified by the leverage at the handlebar ends. Cross-check by applying the front brake and pulling and pushing the machine backward and forward.

2 As the bottom bearing diameter is an integral part of the stanchion if there is any appreciable wear on this diameter or if there are any score marks on the stanchion then it must be renewed.

3 The bottom bearing diameter of the stanchion slides inside the fork leg. Check the inside of the fork leg and the top bearing bush for wear or score marks and replace if necessary. It is not practical to hone out the fork legs or bushes as oversize stanchions are not available.

7 Steering head bearings - examination and replacement

1 Before commencing to reassemble the forks, inspect the steering head races. The ball bearing tracks should be polished and free from indentations and cracks. If signs of wear or damage are evident, the cups and cones must be replaced. They are a tight press fit and need to be drifted out of position.

2 Ball bearings are cheap. Each race contains nineteen ¼ inch balls which should be replaced without question if the originals are marked or discoloured. To hold the ball bearings in place whilst the forks are re-attached, pack the bearings with grease.

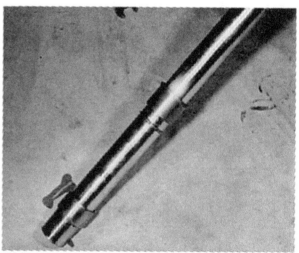

6.1 Arrangement of bushes on stanchion

8 Front forks - reassembly

1 To reassemble the front forks, follow the dismantling procedure in reverse. Extreme care should be taken when assembling the plated·cup and oil seal on the stanchion as the seal is easily damaged. If a small plastic bag is placed over the end of the inner tube this will avoid damage to the seal and facilitate the assembly. It is advisable to smear the sliding members with grease as well as the inside lips of each seal.

2 Tighten the steering head carefully, so that all play is eliminated without placing undue stress on the bearings. The adjustment is correct if all play is eliminated and the handlebars will swing to full lock of their own accord when given a push on one end.

3 It is possible to place several tons pressure on the steering head bearings if they are overtightened. The usual symptom of overtight bearings is a tendency for the machine to roll at low speeds, even though the handlebars may appear to turn quite freely.

4 If, after assembly, it is found that the forks are incorrectly aligned or unduly stiff in action, loosen the front wheel spindle, the two top fork leg nuts and the pinch bolts in both the top and bottom yokes. The forks should then be pumped up and down several times to re-align them. Retighten all the nuts and bolts in the same order, finishing with the steering head pinch bolt.

5 This same procedure can be adopted if the forks are misaligned after an accident. Often the legs will twist within the fork yokes giving the impression of more serious damage, even though no structural damage has occurred.

6 Do not forget to add 154 cc of SAE 10W/30 SE engine oil or SAE10 fork oil to the right-hand fork leg and 136 cc to the left-hand fork leg before replacing the bolts and washers at the top of the fork legs.

9 Frame assembly - examination and renovation

1 The frame is unlikely to need any special attention unless the machine has been involved in an accident or has covered a very large mileage. Small welding and straightening jobs are possible, but care must be taken to limit the amount of heat used and the area to be heated because the load carrying properties diminish when the metal is heated excessively.

2 Frame alignment should be checked when the machine is complete. Fig. 5.3 shows how a board placed each side of the rear wheel can be used as a guide to alignment. It is, of course, necessary to ensure that both wheels are centrally disposed within their respective forks before carrying out this check.

3 Serious damage is not repairable because the frame is made from steel pressings. The purchase of a new frame is invariably cheaper than the cost of attempting to straighten a damaged frame, especially when the necessary jigs for correct alignment are not available.

10 Swinging arm rear suspension - removal from frame

1 After an extended period of service the Silentbloc type swinging arm bushes will wear and need replacing. The rear suspension units are of a sealed type and if they do not function properly cannot be repaired but must be renewed as a matched pair.

2 Commence operations by placing the machine on the centre stand.

3 Remove the rear brake adjuster nut and slide the brake rod out of the brake arm.

4 Remove the split pin from the rear brake plate anchor bolt and remove the anchor nut and washer. The anchor arm can now be pulled clear.

5 On the left hand side of the machine the small wheel spindle nut is now removed and the wheel spindle pulled out from the

right hand side. The spacer between the wheel and the swinging arm can now be pulled clear.

6 The rear wheel can now be pulled clear of the sprocket and if the machine is tilted to one side the wheel can be pulled clear of the machine.

7 The pedalling components should now be removed as described in Chapter 1, Section 5, paragraphs 5 to 7.

8 The two bolts holding the chainguard and the chainguard itself can now be removed.

9 Disconnect the final drive chain at the spring link and pull it clear of the rear sprocket.

10 Remove the large wheel spindle nut and adjuster to allow the sprocket to be pulled clear of the machine.

11 Undo the acorn nuts at the bottom of the rear suspension units and remove the bolts and washers.

12 The nut on the silencer bracket should be removed and the swinging arm pivot rod pulled clear. The swinging arm is now detached from the machine for further work.

11 Swinging arm rear suspension - renovation and reassembly

1 The Silentbloc type swinging arm bushes can be tapped out of the swinging arm and the new ones tapped in but care should be taken to tap only the outer metal of the bushes or damage to the rubber and its bonding will result. As an alternative, the new bushes can be used to remove the old ones provided that they are pressed in.

2 To reassemble the swinging arm, reverse the removal procedure ensuring that the chain and rear brake are properly adjusted.

12 Rear suspension units - examination

These are sealed units and cannot be repaired. The units should be checked for any oil leaks and the rubber bushes in each end checked to ensure that they have not perished.

13 Centre stand - examination

1 The centre stand is attached to a lug on the bottom of the frame, to provide a convenient means of parking the machine on level ground. It pivots on a long bolt which passes through the lug, secured by a nut and washer. A return spring retracts the stand when the machine is pushed forward, so that it can be wheeled prior to riding.

2 The condition of the return spring and the return action should be checked regularly, also the security of the retaining nut and bolt. If the stand stops whilst the machine is in motion it could catch in some obstacle in the road and unseat the rider.

14 Speedometer - removal and replacement

1 A speedometer of the magnetic type is fitted to a support bar above the headlight. It contains also the odometer for recording the total mileage covered by the machine.

2 The speedometer is retained by two clips. To remove the speedometer, detach the drive cable, disconnect the electrical wiring and remove the two clips under the mounting bracket ensuring that the D-shaped washers are not lost. The speedometer can now be removed with its rubber cup.

3 Although a speedometer on a machine of less than 100 cc capacity is not a statutory requirement in the UK, if one is fitted it must be in good working order. Reference to the mileage reading shown on the odometer is a good way of keeping in pace with the routine maintenance schedule.

4 Apart from defects in either the speedometer drive or in the drive cable itself, a speedometer which malfunctions is difficult to repair. Fit a replacement or alternatively, entrust the repair to an instrument repair specialist.

10.3 Remove adjuster nut from rear brake rod

10.4 Disconnect anchor arm from rear brake plate

10.6a Pull out spindle after removing nut on left-hand end and ...

10.6b ... tilt machine slightly to give clearance for wheel removal

10.8 Remove final drive chainguard

10.12 After nut on silencer bracket is removed, pivot rod will pull out from left

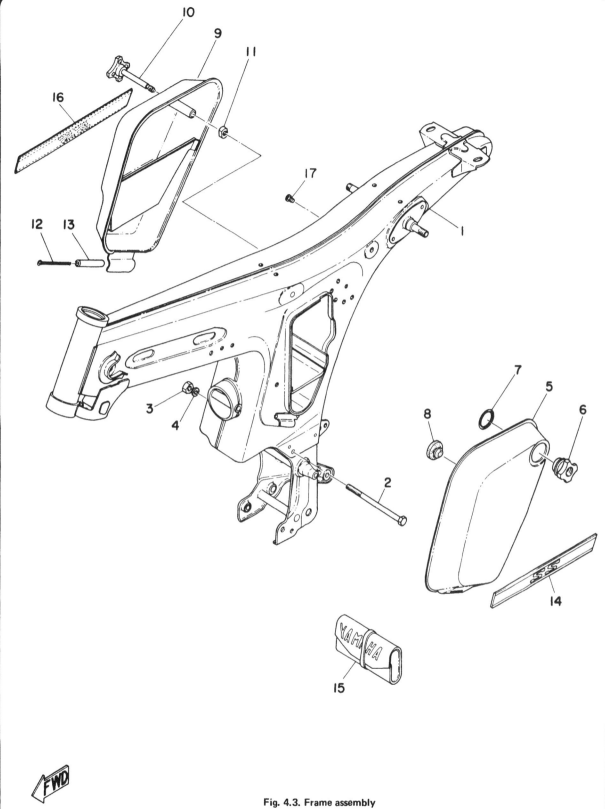

Fig. 4.3. Frame assembly

1 Frame complete	6 Knob for left-hand	10 Knob for right-hand	14 Side cover transfer
2 Bolt	cover	cover	15 Tool roll
3 Nut	7 Limit ring	11 Nut for latch clamp	16 Side cover transfer
4 Spring washer	8 Rubber for side cover	12 Split pin	17 Blind plug
5 Left-hand side cover	9 Right-hand side cover	13 Bush	

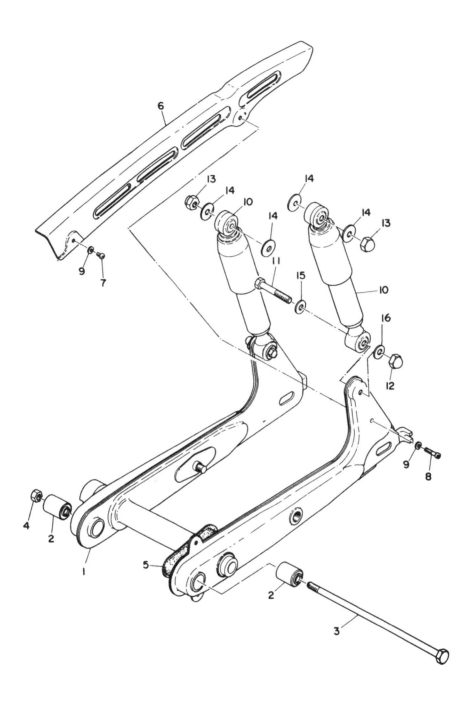

Fig. 4.4. Swinging arm fork and suspension units

1 Swinging arm fork complete
2 Silentbloc bush for swinging arm pivot - 2 off
3 Pivot for swinging arm fork

4 Nut for pivot
5 Guard seal
6 Chainguard
7 Panhead screw for chainguard (front)

8 Panhead screw for chainguard (front)
9 Spring washer - 2 off
10 Rear suspension unit - 2 off
11 Bolt for rear suspension unit mounting - 2 off

12 Acorn nut - 2 off
13 Acorn nut - 2 off
14 Washer - 4 off
15 Plain washer - 2 off
16 Plain washer - 2 off

15 Speedometer cable - inspection and maintenance

1 It is advisable to detach the speedometer drive cable from time to time, in order to check whether it is adequately lubricated and whether the outer covering is compressed or damaged at any point along its run. A jerky or sluggish speedometer movement can often be attributed to a cable fault.

2 To grease the cable, detach the drive from the speedometer head and withdraw the inner cable. After removing the old grease, clean the cable with a petrol soaked rag and examine the cable for broken strands or other damage.

3 Regrease the cable with high melting point grease, taking care not to grease the last six inches at the end where the cable enters the speedometer head. If this precaution is not observed, grease will work into the speedometer head and immobilise the movement.

4 If the speedometer and the odometer stop working, it is probable that the inner cable has broken. Inspection will show the cause of the trouble; if the inner cable has broken it can be renewed on its own and reinserted in the outer covering, after greasing. Never fit a new inner cable alone if the outer covering is damaged also or is compressed at any point along its run.

16 Dualseat - removal

The dualseat is used to hold the petrol tank on the machine and is fixed by two bolts at the front and two nuts at the rear.

After the removal of these fasteners, the dualseat simply lifts off the machine.

17 Steering head lock

All models are fitted with a steering head lock. When the lock is actuated, a tongue protrudes through a hole in an extension of the steering head base, to secure the handlebars on full lock. No attention is necessary other than the occasional application of light oil. If the lock malfunctions, it must be replaced.

18 Cleaning - general

1 After removing all surface dirt with a rag or sponge which is washed frequently in clean water, the application of car polish or wax will give a good finish to the cycle parts of the machine. The plated parts should require only a wipe over with a damp rag, unless salt has caused heavy corrosion. Under these circumstances one of the proprietary chromium plating cleaners can be used.

2 If possible, the machine should be wiped over immediately after it has been used in the wet, so that it is not garaged in damp conditions which will promote rusting. Make sure the chain is wiped and if necessary re-oil it, to prevent water from entering the rollers and causing harshness with an accompanying greater rate of wear. Remember there is less chance of water entering the control cables if they are lubricated regularly, as recommended in the Routine Maintenance Section.

19 Fault diagnosis - frame and forks

Symptom	Reason/s	Remedy
Machine is unduly sensitive to road surface irregularities	Fork and/or rear suspension units damping ineffective	Check oil level in forks. Renew suspension units as a pair.
Machine rolls at low speeds	Steering head bearings overtight or damaged	Slacken bearing adjustment. If no improvement, dismantle and inspect head bearings.
Machine tends to wander; steering is imprecise	Worn swinging arm suspension bearings	Check and if necessary renew pivot spindle and bush.
Fork action stiff	Fork legs have twisted in yokes or have been drawn together at lower ends	Slacken off spindle nut, pinch bolts in yoke and fork top nuts. Pump forks several times before re-tightening from bottom.
Forks judder when front brake is applied	Worn fork bushes Steering head bearings too slack	Strip forks and renew worn bushes. Re-adjust to take up play.
Wheels seem out of alignment	Frame distorted as result of accident damage	Check frame. If bent, specialist repair is necessary.

Chapter 5 Wheels, brakes and tyres

Refer to Chapter 7 for information relating to the 1975 on models

Contents

Specifications

Wheels:

Size ... ... 17 in. diameter, front and rear. Not interchangeable

Tyres:

Size ... ... 17 in. x 2.25 in. front
17 in. x 2.50 in. rear

Tyre pressures:

Front ... ... 20 p.s.i. (1.4 kg/cm^2)

Rear ... ... 28 psi (2.0 kg/cm^2)

Brakes:

Diameter of brake drum ... ... 4.33 in. (110 mm) front and rear

1 General description

The wheels are 17 inch diameter with a larger section tyre on the rear being 2.50 inch as opposed to 2.25 inch on the front. Each hub contains a 110 mm internally expanding brake which gives excellent braking. Both wheels are of the quickly-detachable type but are not interchangeable as the rear hub contains the rubber transmission shock absorber.

2 Front wheel - inspection and renovation

1 Place the machine on the centre stand so that the front wheel is raised clear of the ground. Spin the wheel and check for rim alignment or run-out. Small irregularities can be corrected by tightening the spokes in the area affected, although a certain amount of experience is advisable if over-correction is to be avoided.
2 Any flats in the wheel rim should be evident at the same time. These are much more difficult to remove and in most cases the wheel will need to be rebuilt on a new rim. Apart from the

effect on stability, there is greater risk of damage to the tyre bead and walls if the machine is run with a deformed wheel. In an extreme case the tyre can even separate from the rim.
3 Check for loose or broken spokes. Tapping the spokes is the best guide to tension. A loose spoke will produce a quite different sound and should be tightened by turning the nipple in an anticlockwise direction. Always recheck for run-out by spinning the wheel again.
4 If it is necessary to turn a spoke nipple an excessive amount to restore tension, it is advisable to remove the tyre and tube so that the end of the spoke that now protrudes into the wheel rim can be ground flush. If this precaution is not taken, there is danger of the spoke end chafing the inner tube and causing an eventual puncture.

3 Front wheel - removal

1 Commence operations by placing the machine on the centre stand.

2 Slacken the front brake cable and disconnect it from the brake.

3 Remove the circlip retaining the speedometer cable in the brake plate and pull the cable clear.

4 Undo the front wheel spindle nut and withdraw the wheel spindle. Support the machine to stop it from toppling forward.

5 The front wheel will now drop clear of the machine. Remove the spacer on the left hand side of the wheel to avoid it being lost.

4 Front brake assembly - inspection, renovation and reassembly

1 To remove the brake assembly, lift it out from the brake drum.

2 Examine the brake linings. If they are wearing thin or unevenly, the brake shoes should be replaced. The linings are bonded on and cannot be replaced as a separate item.

3 To remove the brake shoes from the brake plate assembly, arrange the operating lever so that the brakes are in the 'full on' position and then pull the shoes apart whilst lifting them upward in the form of a 'V'. When they are clear of the brake plate, the return springs can be removed and the shoes separated.

4 Before replacing the brake shoes, check that the brake operating cam is working smoothly and is not binding in its pivot. The cam can be removed by withdrawing the retaining nut on the operating arm and pulling the arm off the shaft. Before removing the arm, it is advisable to mark its position in relation to the shaft, so that it can be relocated correctly. The shaft should be greased prior to reassembly and also a light smear of grease placed on the faces of the operating cam.

5 Check the inner surface of the brake drum on which the brake shoes bear. The surface should be smooth and free from score marks or indentations, otherwise reduced braking efficiency will be inevitable. Remove all traces of brake lining dust and wipe with a clean rag soaked in petrol to remove any traces of grease or oil.

6 If the brake drum has become scored, specialist attention is required. It is possible to skim a brake drum in a lathe provided the score marks are not too deep. Under these circumstances, packing will have to be added to the ends of the brake shoes, to compensate for the amount of metal removed from the surface of the drum.

7 To reassemble the brake shoes on the brake plate, fit the return springs first and then force the shoes apart, holding them in a 'V' formation. If they are now located with the brake operating cam and pivot they can usually be snapped into

position by pressing downward. Never use excessive force, otherwise there is risk of distorting the shoes permanently.

5 Front wheel bearings - inspection and replacement

1 The front wheel bearings are of the ball journal type and are not adjustable. If the bearings are worn, indicated by side play on the wheel rim, the bearings must be renewed.

2 Access to the wheel bearings is gained when the brake assembly has been removed from the front wheel.

3 To remove the first bearing drive the bearing spacer out from the brake hub side using a suitable drift. This will push out the bearing and the oil seal. The second bearing can then be driven out from the other side of the wheel.

4 To reassemble the bearings, drive one side in first, fit the bearing spacer and drive the second bearing in. Refit the oil seal taking care not to damage it.

6 Speedometer drive gears - examination and replacement

1 The drive gears should be checked for wear or broken teeth and renewed if necessary.

2 To renew the large drive gear on the hub, it may be necessary to heat the drive ring before driving the hub out from the inside. The oil seal behind the gear should also be renewed as it will have been damaged. The new drive gear may also need heating before fitting as the gear is a very tight fit on the hub. Extreme caution should be used to avoid damage to the new oil seal.

3 To renew the small worm gear in the brake plate it is necessary to prise out the small oil seal, unscrew the bush and withdraw the worm gear. Reassemble in reverse, using the above procedure.

4 Thoroughly grease the gears before refitting the brake plate.

7 Front wheel - replacement

1 To replace the front wheel, reverse the removal procedure and ensure that the peg on the forks locates in the slot in the brake plate.

2 Reconnect the front brake and check that the brake functions correctly, especially if the adjustment has been altered or the brake operating arm has been removed and replaced during the dismantling operation.

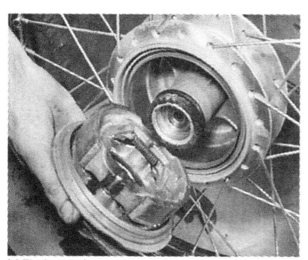

4.1 The front brake assembly will lift from brake plate

4.2 Examine brake linings for wear or damage

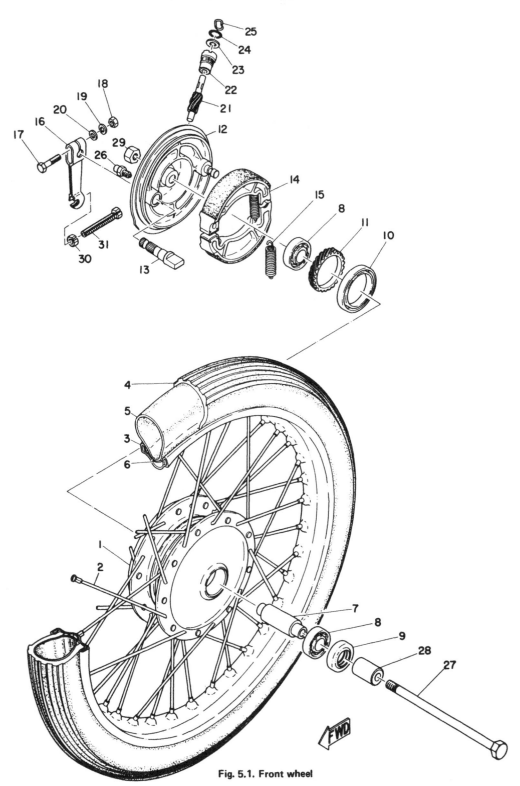

Fig. 5.1. Front wheel

1	Hub	7	Wheel bearing spacer
2	Spoke set	8	Wheel bearings - 2 off
3	Rim (17 inch diameter)	9	Left-hand oil seal
4	Front tyre 2.25 x 17 inch	10	Right-hand oil seal
5	Inner tube 2.25 x 17 inch	11	Speedometer drive pinion
6	Rim tape	12	Brake plate

1 Hub
2 Spoke set
3 Rim (17 inch diameter)
4 Front tyre 2.25 x 17 inch
5 Inner tube 2.25 x 17 inch
6 Rim tape

7 Wheel bearing spacer
8 Wheel bearings - 2 off
9 Left-hand oil seal
10 Right-hand oil seal
11 Speedometer drive pinion
12 Brake plate
13 Brake operating cam
14 Brake shoe complete with lining - 2 off

15 Brake shoe return spring - 2 off
16 Brake operating arm
17 Bolt for brake operating arm
18 Nut
19 Spring washer
20 Plain washer
21 Speedometer driven pinion

22 Bush
23 Oil seal
24 'O' ring
25 Stop ring (clip)
26 Grease nipple
27 Front wheel spindle
28 Spacer
29 Wheel spindle nut
30 Brake cable adjuster
31 Adjuster locknut

8 Rear wheel - inspection and renovation

1 Place the machine on the centre stand so that the rear wheel is clear of the ground. Check the wheel for rim alignment, damage to the rim or loose or broken spokes, by following the procedure adopted for the front wheel in the preceding Section.
2 Note that although the front and rear wheels are of identical size and have hubs of similar diameter, they cannot be interchanged.

9 Rear wheel - removal

1 Commence operations by placing the machine on the centre stand.
2 Remove the rear brake adjuster nut and slide the brake rod out of the brake arm.

3 Remove the split pin from the brake plate anchor bolt and remove the anchor nut and washer. The anchor arm can now be pulled clear.
4 On the left hand side of the machine the small wheel spindle nut is now removed and the wheel spindle pulled out from the right hand side. The spacer between the wheel and the swinging arm can now be pulled clear.
5 The rear wheel can now be pulled clear of the sprocket and if the machine is tilted to one side the wheel can be pulled clear of the machine.

10 Rear brake assembly - inspection, renovation and reassembly

The rear brake assembly is removed and dismantled by following the procedure adopted for the front brake assembly, as detailed in Section 4 of this Chapter.

6.1 Check speedometer drive pinions for wear

6.2 Large drive pinion is shrink fit on hub

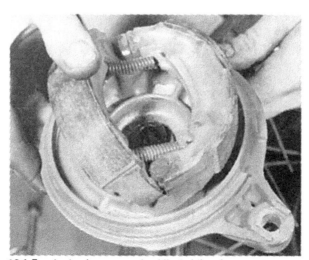
10.1 Ease brake shoes upwards whilst holding them apart

11 Rear wheel bearings - inspection and replacement

1 The rear wheel bearings are of the ball journal type and are not adjustable. If the bearings are worn, indicated by side play on the wheel rim, the bearings must be renewed.

2 To remove the first bearing drive the bearing spacer out from the sprocket side using a suitable drift. This will push out the bearing and the oil seal. The second bearing can then be driven out from the other side of the wheel.

3 To reassemble the bearings, drive one side in first, fit the bearing spacer and drive the second bearing in. Refit the oil seal taking care not to damage it.

4 It is advisable to replace the O ring on the sprocket side of the hub to ensure that no grease can reach the shock absorber rubbers.

12 Rear wheel sprocket and shock absorber assembly - removal, examination and replacement

1 The rear wheel sprocket assembly is removed as follows:

Disconnect the chain at the spring link and pull it clear of the rear sprocket. Remove the large wheel spindle nut and adjuster to allow the sprocket assembly to be pulled clear of the machine. (It is assumed that the rear wheel has already been removed.)

2 The sprocket assembly has a ball journal bearing and if worn the following procedure should be followed:

3 Remove the stub axle and spacer from the assembly. Prise out the oil seal. Remove the retaining circlip and drive out the bearing.

4 To replace the bearing, reverse the procedure using a new oil seal as the old one will be damaged when it is removed.

5 It is unlikely that the sprocket will require renewal until a very substantial mileage has been covered. The usual signs of wear occur when the teeth assume a hooked or very shallow formation which will cause rapid wear of the chain. A worn sprocket must be replaced, together with the gearbox final drive sprocket and the chain. Always replace the final drive assembly as a complete set, otherwise rapid wear will occur as the result of running old and new parts together.

6 To replace the sprocket, prise down the tab washers and remove the four retaining bolts. Reassemble in the reverse order.

11.2a Oil seal is in front of wheel bearings, will be displaced before ...

11.2b ... bearing itself emerges from hub

11.2c Remaining bearing will drive out from opposite direction

11.3 When refitting, do not omit bearing spacer

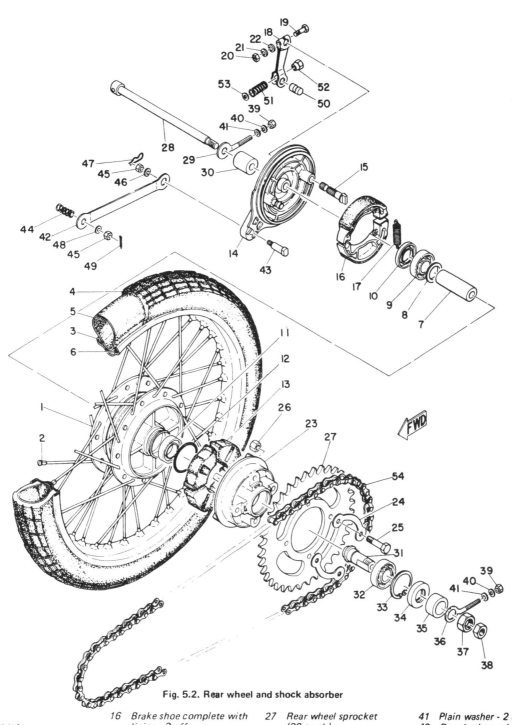

Fig. 5.2. Rear wheel and shock absorber

1	Hub	16	Brake shoe complete with lining - 2 off	27	Rear wheel sprocket (39 teeth)	41	Plain washer - 2 off
2	Spoke set	17	Brake shoe return spring - 2 off	28	Rear wheel spindle	42	Rear brake anchor arm
3	Rim (17 inch diameter)	18	Brake operating arm	29	Right-hand chain adjuster	43	Anchor arm bolt
4	Rear tyre 2.50 x 17 inch	19	Bolt for brake operating arm	30	Wheel spacer	44	Anchor arm spring
5	Inner tube 2.50 x 17 inch	20	Nut	31	Stub axle	45	Nut - 2 off
6	Rim tape	21	Spring washer	32	Stub axle bearing	46	Spring washer
7	Wheel bearing spacer	22	Plain washer	33	Circlip	47	Clip for anchor arm bolt
8	Spacer flange	23	Shock absorber hub (cush drive)	34	Oil seal	48	Plain washer
9	Right-hand wheel bearing	24	Lock washer for sprocket bolts - 2 off	35	Distance piece	49	Split pin
10	Right-hand oil seal	25	Sprocket bolt - 4 off	36	Left-hand chain adjuster	50	Trunnion
11	Left-hand wheel bearing	26	Nut - 4 off	37	Stub axle nut	51	Spring for brake operating rod
12	'O' ring			38	Wheel spindle nut	52	Brake adjusting nut
13	Shock absorber rubbers - 4 off			39	Chain adjuster nut - 2 off	53	Plain washer
14	Rear brake plate			40	Spring washer - 2 off	54	Final drive chain (96 links)
15	Brake operating cam						

7   The shock absorber rubbers will remain in the wheel hub and should be checked for any damage or deterioration. All oil or grease should be wiped away as this may cause premature deterioration.

13 Rear wheel - reassembly

1   To refit the rear wheel reverse the removal procedure.
2   Before fully tightening all the nuts ensure that the final chain tension and the brake adjustment are correct.
3   Check also whether the wheel alignment is correct. The rear chainstays are marked with indentations so that a visual check can be made.

14 Pedal assembly - examination and replacement

1   As pedals are fitted to this machine to make it legal for 16 year olds to ride, there should be very little wear and parts should only need replacing due to accidental damage.
2   The pedals are normal bicycle pedals which are unscrewed and new ones screwed in if necessary. Note that the right hand pedal has a left hand thread.
3   If the left hand crank and chainwheel need renewing, the procedure is found in Chapter 1, Section 5, paragraphs 5 to 7.
·4   If the right hand crank needs renewing the rubber boot should be pulled up the crank to expose a circlip on the shaft, which retains the crank. There is a slot in the crank into which the peg in the shaft must go to allow the crank to slide off.
5   To replace the spring it is easiest to remove the shaft from the machine prior to detaching the small circlip and spring.
6   If either crank is bent, renewal is advised. They are difficult to straighten without risk of a fracture occurring.

15 Front and rear brakes - adjustment

1   The front brake adjuster is located on the front brake plate. The brake should be adjusted so that the wheel is free to revolve before pressure is applied to the handlebar lever and is applied fully before the handlebar lever touches the handlebar. Make sure that the adjuster locknut is tight after the correct adjustment has been made.

12.1a Rear wheel sprocket bolts to shock absorber plate ...

12.1b ... contains stub axle and own bearing

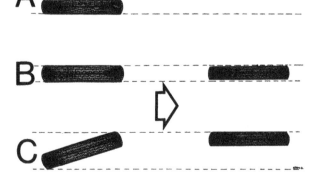

Fig. 5.3. Method of checking wheel alignment

A and C incorrect   *B correct*

13.2 Don't omit spring clip after final tightening

Tyre removal: Deflate inner tube and insert lever in close proximity to tyre valve

Use two levers to work bead over the edge of rim

When first bead is clear, remove tyre as shown

Tyre fitting: Inflate inner tube and insert in tyre

Lay tyre on rim and feed valve through hole in rim

Work first bead over rim, using lever in final section

Use similar technique for second bead, finish at tyre valve position

Push valve and tube up into tyre when fitting final section, to avoid trapping

2   The rear brake is adjusted by means of the adjusting nut on the end of the brake operating rod. This nut is self-locking. Adjustment is largely a matter of personal choice, but excessive travel of the footbrake pedal should not be necessary before the brake is applied fully.

3   **Check frequently that the rear brake torque arm bolt is tight. If the torque arm becomes detached, the rear brake will lock in the full-on position immediately it is applied and may give rise to a serious accident.**

4   Efficient brakes depend on good leverage of the operating arms. The angle between the brake operating arm and the cable or rod should never exceed 90° when the brake is fully applied.

16   Final drive chain - inspection and lubrication

1   Periodically, the tension of the final drive chain should be checked by measuring the amount of play in the middle of the bottom run. The chain is in correct adjustment if there is from 5/8 to 7/8 inch play at the tightest spot along its length, when the machine is standing upright on its wheels.

2   To adjust the chain, slacken the rear wheel nuts and draw the rear wheel spindle either forward or backward until the correct tension is achieved, by means of the chain adjusters. Tighten the wheel nuts and check again that the chain tension is correct.

3   Always adjust the draw bolts an identical amount, otherwise the rear wheel will be thrown out of alignment. If in doubt about the correctness of wheel alignment, refer to Fig. 5.3.

4   After a period of running, the chain will require lubrication. Lack of oil will accelerate the rate of wear of both chain and sprockets, leading to harsh transmission. The application of an engine oil from an oil can still serves as a satisfactory lubricant, but it is preferable to remove the chain at regular intervals and immerse it in a molten lubricant such as Linklyfe, after it has been cleaned in a paraffin bath. This latter type of lubricant achieves better penetration of the chain links and rollers and is less likely to be thrown off when the chain is in motion.

5   To check whether the chain is due for replacement, lay it lengthwise in a straight line and compress it so that all play is taken up. Anchor one end and then pull on the other end to take up the play in the opposite direction. If the chain extends by more than the distance between two adjacent rollers, it should be replaced in conjunction with the sprockets. Note that this check should be made after the chain has been washed, but before the lubricant has been applied, otherwise the lubricant may take up some of the play.

6   When replacing the chain, make sure the spring link is positioned correctly, with the closed end facing the direction of travel. Reconnection is made easier if the ends of the chain are pressed into the teeth of the rear wheel sprocket.

17   Pedalling gear chain - inspection and lubrication

1   The same advice given in the preceding Section applies to the pedalling gear chain, although this chain is only in occasional use and will not be subjected to anything like the same amount of wear.

2   The pedalling gear chain does not require checking for tension because it is not adjustable. If it is badly worn, it must be renewed.

18   Tyres - removal and replacement

1   At some time or other the need will arise to remove and replace the tyres, either as the result of a puncture or because a replacement is required to offset wear. To the inexperienced, tyre changing represents a formidable task yet if a few simple rules are observed and the technique learned, the whole operation is surprisingly simple.

2   To remove the tyre from either wheel, first detach the wheel **from the machine** by following the procedure in Chapters 5.3 or

17.2 Pedalling chain has no means of adjustment

5.9 depending on whether the front or the rear wheel is involved. Deflate the tyre by removing the valve insert and when it is fully deflated, push the bead of the tyre away from the wheel rim on both sides so that the bead enters the centre well of the rim. Remove the locking cap and push the tyre valve into the tyre itself.

3   Insert a tyre lever close to the valve and lever the edge of the tyre over the outside of the wheel rim. Very little force should be necessary; if resistance is encountered it is probably due to the fact that the tyre beads have not entered the well of the wheel rim all the way round the tyre.

4   Once the tyre has been edged over the wheel rim, it is easy to work around the wheel rim so that the tyre is completely free on one side. At this stage, the inner tube can be removed.

5   Working from the other side of the wheel, ease the other edge of the tyre over the outside of the wheel rim that is furthest away. Continue to work around the rim until the tyre is free completely from the rim.

6   If a puncture has necessitated the removal of the tyre, re-inflate the inner tube and immerse it in a bowl of water to trace the source of the leak. Mark its position and deflate the tube. Dry the tube and clean the area around the puncture with a petrol soaked rag. When the surface has dried, apply the rubber solution and allow this to dry before removing the backing from the patch and applying the patch to the surface.

7   It is best to use a patch of the self vulcanising type, which will form a very permanent repair. Note that it may be necessary to remove a protective covering from the top surface of the patch, after it has sealed in position. Inner tubes made from synthetic rubber may require a special type of patch and adhesive, if a satisfactory bond is to be achieved.

8   Before replacing the tyre, check the inside to make sure the agent that caused the puncture is not trapped. Check also the outside of the tyre, particularly the tread area, to make sure nothing is trapped that may cause a further puncture.

9   If the inner tube has been patched on a number of past occasions, or if there is a tear or large hole, it is preferable to discard it and fit a replacement. Sudden deflation may cause an accident, particularly if it occurs with the front wheel.

10   To replace the tyre, inflate the inner tube sufficiently for it to assume a circular shape but only just. Then push it into the tyre so that it is enclosed completely. Lay the tyre on the wheel at an angle and insert the valve captive in its correct location.

11   Starting at the point furthest from the valve, push the tyre bead over the edge of the wheel rim until it is located in the central well. Continue to work around the tyre in this fashion until the whole of one side of the tyre is on the rim. It may be necessary to use a tyre lever during the final stages.

12   Make sure there is no pull on the tyre valve and again

commencing with the area furthest from the valve, ease the other bead of the tyre over the edge of the rim. Finish with the area close to the valve, pushing the valve up into the tyre until the locking cap touches the rim. This will ensure the inner tube is not trapped when the last section of the bead is edged over the rim with a tyre lever.

13 Check that the inner tube is not trapped at any point. Re-inflate the inner tube, and check that the tyre is seating correctly around the wheel rim. There should be a thin rib moulded around the wall of the tyre on both sides, which should be equidistant from the wheel rim at all points. If the tyre is unevenly located on the rim, try bouncing the wheel when the tyre is at the recommended pressure. It is probable that one of the beads has not pulled clear of the centre well.

14 Always run the tyres at the recommended pressures and never under or over inflate. The correct pressures for solo use are given in the Specifications Section of this Chapter. If a pillion passenger is carried, increase the rear tyre pressure only by approximately 4 psi.

15 Tyre replacement is aided by dusting the side walls, particularly in the vicinity of the beads, with a liberal coating of French chalk. Washing up liquid can also be used to good effect, but this has the disadvantage of causing the inner surfaces of the wheel rim to rust.

16 Never replace the inner tube and tyre without the rim tape in position. If this precaution is overlooked there is good chance of the ends of the spoke nipples chafing the inner tube and causing a crop of punctures.

17 Never fit a tyre that has a damaged tread or side walls. Apart from the legal aspects, there is a very great risk of a blow-out, which can have serious consequences on any two-wheel vehicle.

18 Tyre valves rarely give trouble, but it is always advisable to check whether the valve itself is leaking before removing the tyre. Do not forget to fit the dust cap which forms an effective second seal.

19 Fault diagnosis - wheels, brakes and tyres

Symptom	Reason/s	Remedy
Handlebars oscillate at low speeds	Buckle or flat in wheel rim, most probably front wheel	Check rim alignment by spinning wheel. Correct by retensioning spokes or having wheel rebuilt on new rim.
	Tyre not straight on rim	Check tyre alignment.
Machine lacks power and accelerates poorly	Brakes binding	Warm brake drums provide best evidence. Readjust brakes.
Brakes grab when applied gently	Ends of brake shoes not chamfered	Chamfer with file.
	Elliptical brake drum	Lightly skim in lathe (specialist attention needed).
Brake pull-off sluggish	Brake cam binding in housing	Free and grease.
	Weak brake shoe springs	Replace, if brake springs not displaced.
Harsh transmission	Worn or badly adjusted chains	Adjust or replace as necessary.
	Hooked or badly worn sprockets	Replace as a pair, together with chain.

Chapter 6 Electrical system

Refer to Chapter 7 for information relating to the 1975 on models

Contents

Specifications

Flywheel generator:

Make	Mitsubishi
Type	FAZ IQL or FII - L40
Output lighting winding	8.7 volts (4 amp AC)
Neutral switch	Asaki YNS type
Main switch	Asaki
Stop lamp switch	Asaki YS10

Battery	YUASA BST 2 - 6
Silicon rectifier	Feyi COZ - H 1/1
	YUASA SZ - 3A
Horn	Imasen SM3 - 6V
Flasher relay	Showa 8 - 9
Fuse rating	10 amp

Bulbs:

Headlight (Koito)	7 volts	18/18 watt
Tail light (Imasen)	6 volts	21/5 watt
Speedometer (Nippon Seiki)	Neutral light	6 volts	3 watt
	Meter light	6 volts	1.5 watt

1 General description

The flywheel generator fitted to the Yamaha produces an alternating current and has a centre tapping on the generator coil as a means of varying the charge rate. When the engine is running without the lights, only half the generator coil is used to produce power, which is rectified before being fed into the battery. The battery then supplies the power for all the ancillary equipment, ie the horn, neutral light, stop light and flashers.

When the main lights are switched on the full generator coil is used and alternating current is fed directly to the lights with a switched connection to the rectifier. This latter arrangement keeps power supplied to the battery and avoids blowing the tail and speedometer bulbs if the headlamp bulb fails when changing from full to dipped headlamp.

2 Flywheel generator - checking output

As explained in Chapter 3, the output can be checked only with specialised test equipment of the multi-meter type. If the generator is suspect, it should be checked by either a Yamaha agent or an auto-electrical expert.

3 Battery - examination and maintenance

1 Maintenance is normally limited to keeping the electrolyte

level just above the plates and separators. Modern batteries have translucent plastics cases, which make the check of electrolyte level much easier.

2 Unless acid is spilt, which may occur if the machine falls over, the electrolyte should always be topped up with distilled water until the correct level is restored. If acid is spilt on any part of the machine, it should be neutralised with an alkali such as washing soda and washed away with plenty of water, otherwise serious corrosion will occur. Top up with sulphuric acid of the correct specific gravity (1.260 to 1.280) ONLY when spillage has occurred.

3 It is seldom practicable to repair a cracked battery case because the acid already in the joint will prevent the formation of an effective seal. It is always best to replace a cracked battery, especially in view of the corrosion that will be caused by the leakage of acid.

4 Battery - charging procedure

1 Whilst the machine is running, it is unlikely that the battery will require attention other than routine maintenance because the generator will keep it charged. However, if the machine is used for a succession of short journeys only, it is possible that the output from the generator will not be able to keep pace with the heavy electrical demand. Under these circumstances it will be necessary to remove the battery from time to time to have it charged independently.

2 The normal charging rate is 1 amp. A more rapid charge can be given in an emergency, but this should be avoided if possible because it will shorten the working life of the battery.

3 When the battery has been removed from a machine that has been laid up, a 'refresher' charge should be given every six weeks if the battery is to be maintained in good condition.

5 Silicon rectifier - general description

1 The function of the silicon rectifier is to convert the AC produced by the generator to DC so that it can be used to charge the battery and operate the lighting circuit etc. The usual symptom of a defective rectifier is a battery which discharges rapidly because it is receiving no charge from the generator.

2 The rectifier is located where it is not exposed to water or battery acid, which will cause it to malfunction. The question of access is of relatively little importance because the rectifier is unlikely to give trouble during normal operating conditions. It is not practicable to repair a damaged rectifier; replacement is the only satisfactory solution. One of the most frequent causes of rectifier failure is the inadvertent connection of the battery in reverse which results in a reverse current flow.

3 Damage to the rectifier is also liable to occur if the machine is run without a battery for any period of time. A high voltage will develop in the absence of any load on the coil which will cause a reverse flow of current and consequent damage to the rectifier cells.

4 It is not possible to check whether the rectifier is functioning correctly without the appropriate test equipment. A Yamaha agent or an auto-electrical expert are best qualified to advise in such cases.

5 Do not loosen the rectifier locking nut or bend, cut, scratch or rotate the wafer. Any such action will cause the electrode alloy coating to peel and destroy the working action.

6 Headlamp - replacing bulbs and adjusting beam height

1 To remove the headlamp rim, unscrew the screw at the base of the rim. The rim will then pull off, complete with the reflector unit and bulbs.

2 The reflector unit contains a double-filament bulb which provides the main and dipped headlamp beams. It is controlled from a dipswitch mounted on the handlebars.

3 It is not necessary to refocus the headlamp when a new bulb is fitted because the bulbs are of the prefocus type. To release the bulb holder from the reflector, twist and pull.

4 Beam height is adjusted by slackening the two headlamp shell retaining nuts and tilting the headlamp either upward or downward. Adjustments should always be made with the rider normally seated.

5 UK lighting regulations stipulate that the lighting system must be arranged to that the light will not dazzle a person standing in the same horizontal plane as the vehicle at a distance greater than 25 yards from the lamp, whose eye level is not less than 3 feet 6 inches above that plane. It is easy to approximate this setting by placing the machine 25 yards away from a wall, on a level road, and setting the beam height so that it is concentrated at the same height as the distance from the centre of the headlamp to the ground. The rider must be seated normally during this operation and also the pillion passenger, if one is carried regularly.

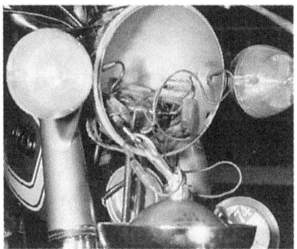

6.2 Reflector unit contains a double filament bulb and no parking light

7 Tail lamp - replacing bulb

1 The moulded plastics cover of the rear lamp is retained by two screws. When these screws are removed, the cover can be removed and the bulb exposed.

2 To release the bulb from its holder, press it inwards, turn to the left and pull out. Press in the new bulb, turn to the right and pull downwards so that the stops locate. Refit the cover, making sure that the rubber moulding which surrounds the cover is located correctly to exclude water.

3 Make sure that the flexible contact makes a good connection with the bottom contact of the bulb, otherwise the bulb may work only intermittently and eventually 'blow'.

4 The tail lamp has a 3 watt rating, and the stop lamp a 10 watt rating.

8 Speedometer bulbs - replacement

1 The speedometer bulb and the neutral warning light bulb holders push into the speedometer head.

2 After the holder is pulled clear the bulbs can be replaced in the same manner as the tail light bulb. The holders are then pushed back into the speedometer head.

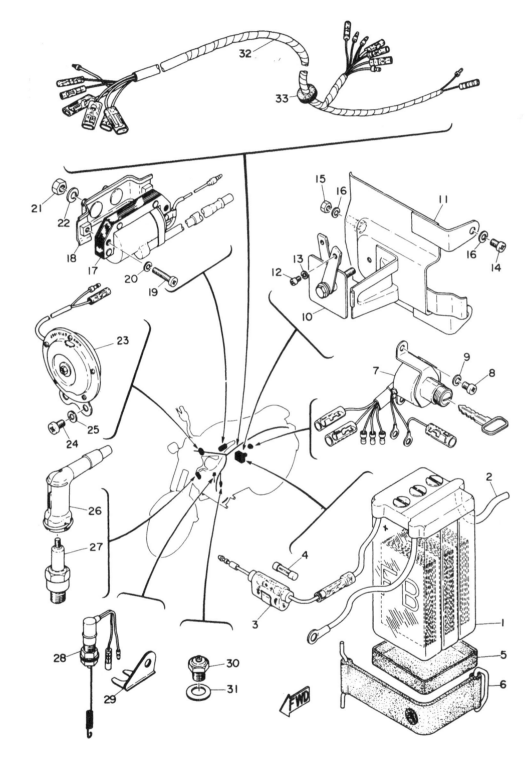

Fig. 6.1. Electrical components

1	Battery		switch	17	Ignition coil	25 Spring washer - 2 off
2	Vent pipe	9	Spring washer	18	Coil holder	26 Plug suppressor cap
3	Fuse holder	10	Rectifier	19	Coil holder screw - 2 off	27 Spark plug
4	Fuse	11	Rectifier mounting plate	20	Spring washer - 2 off	28 Stop lamp switch
5	Battery carrier	12	Panhead screw - 2 off	21	Nut - 2 off	29 Stay for stop lamp switch
6	Battery strap	13	Spring washer - 2 off	22	Spring washer - 2 off	30 Neutral switch
7	Combined ignition and	14	Panhead screw - 2 off	23	Horn	31 Gasket for neutral switch
	lighting switch	15	Nut	24	Screw for horn mounting	32 Wiring harness
8	Panhead screw for retaining	16	Spring washer - 3 off		- 2 off	33 Cable grommet

9 Flasher bulbs - replacement

1 The moulded plastics cover of the flasher lamp is retained by two screws. When these screws are removed, the cover can be removed and the bulb exposed.

2 To release the bulb from its holder, press it inwards, turn to the left and pull out. Press in the new bulb, turn to the right and pull downwards so that the stops locate. Refit the cover making sure the rubber moulding that surrounds the cover is located correctly to exclude water.

10 Horn - location and adjustment

1 The horn is mounted on a bracket attached to the bottom yoke of the forks, immediately behind the front number plate.

2 There is no means of adjusting the horn note.

11 Wiring - layout and inspection

1 The wiring harness is colour-coded and will correspond with the accompanying wiring diagrams.

2 Visual inspection will show whether any breaks or frayed outer coverings are giving rise to short circuits. Another source of trouble may be the snap connectors, particularly where the connector has not been pushed home fully in the outer casing.

3 Intermittent short circuits can often be traced to a chafed wire which passes through or close to a metal component, such as a frame member. Avoid tight bends in the wire or situations where the wire can become trapped or stretched between casings.

12 Ignition and lighting switch

1 The same key operated switch is used for ignition and lighting. This switch is not repairable and if found faulty must be replaced.

2 New keys are not available so if both keys are lost a new switch must be fitted.

3 On no account oil the switch or the oil will spread across the internal contacts and form an effective insulator.

13 Fuse - location and replacement

1 A fuse within a moulded plastics case is incorporated in the electrical system to give protection from a sudden overload, such as may occur during a short circuit. It is found in close proximity to the battery, retained in metal clips. The fuse is rated at 10 amps; a plastic bag containing a spare fuse of similar rating is normally carried in a plastic bag attached to the wiring harness.

2 If a fuse blows, it should not be replaced until a check has shown whether a short circuit has occurred. This will involve checking the electrical circuit to identify and correct the fault. If this precaution is not observed, the replacement fuse, which may be the only spare, may blow immediately on connection.

3 When a fuse blows whilst the machine is running and no spare is available a get you home remedy is to remove the blown fuse and wrap it in silver paper before replacing it in the fuse holder. The silver paper will restore electrical continuity by bridging the broken wire within the fuse. This expedient should never be used if there is evidence of a short circuit or other major electrical fault, otherwise more serious damage will be caused. Replace the 'doctored' fuse at the earliest possible opportunity to restore full circuit protection.

14 Stop lamp switch - adjustment

1 All models have a stop lamp switch fitted to operate in

9.2a Make sure lens sealing gasket is in good order when replacing bulbs

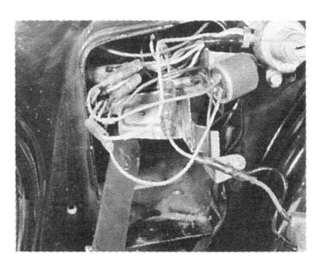

9.2b Flasher unit is located close to ignition/lighting switch

7.4 Tail/stop lamp has twin filament bulb, with offset pins

conjunction with the rear brake pedal. The switch is located immediately to the rear of the crankcase, on the right hand side of the machine. It has a threaded body, permitting a range of adjustment.

2 If the stop lamp is late in operating, slacken the locknuts and turn the body of the lamp in an anticlockwise direction so that the switch rises from the bracket to which it is attached. When

the adjustment seems near correct, tighten the locknuts and test.

3 If the lamp operates too early, the locknuts should be slackened and the switch body turned clockwise so that it is lowered in relation to the mounting bracket.

4 As a guide, the light should operate after the brake pedal has been depressed by about 2 cm (¾ inch).

15 Fault diagnosis - electrical equipment

Symptom	Reason/s	Remedy
Complete electrical failure	Blown fuse	Check wiring and electrical components for short circuit before fitting new 10 amp fuse.
	Isolated battery	Check battery connections, also whether connections show signs of corrosion.
Dim lights, horn inoperative	Discharged battery	Recharge battery with battery charger and check whether alternator is giving correct output.
Constantly 'blowing' bulbs	Vibration, poor earth connection	Check whether bulb holders are secured correctly. Check earth return or connections to frame.

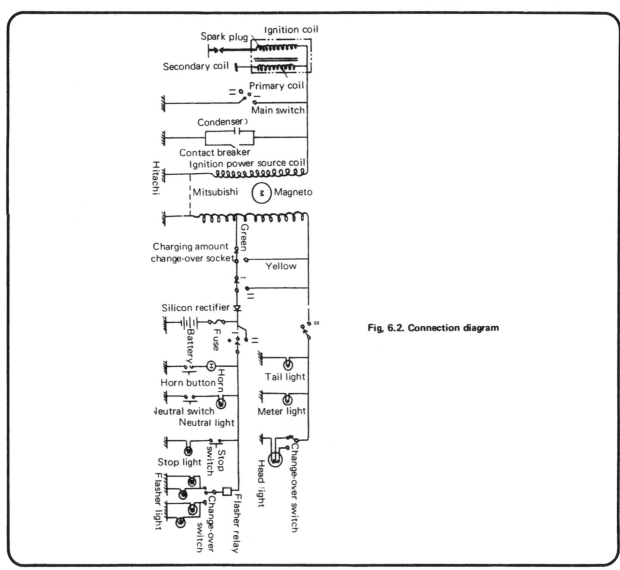

Fig. 6.2. Connection diagram

The FS1/FS1M (2GO model)

The FS1SE (5A1) model

Chapter 7 The 1975 on models

Contents

Specifications

Information is given only where different from, or in addition to, that given in Chapters 1 to 6 of this Manual. Unless otherwise stated, information applies to all models. Models are identified, where necessary, by their Yamaha model codes; refer to Section 1 of this Chapter for details.

Model dimensions and weight

FS1E-DX (596), FS1E-DXA:
Overall width	695 mm (27.4 in)
Dry weight	72 kg (159 lb)

FS1/FS1M (2GO):
Overall length	1780 mm (70.0 in)
Overall width	735 mm (28.9 in)

FS1-DX/FS1M-DX (2GO):
Overall length	1780 mm (70.0 in)
Overall width	695 mm (27.4 in)
Overall height	1010 mm (39.8 in)
Dry weight	71 kg (156 lb)

FS1-DX (3F6), FS1SE, FS1 (2RV, 3AU):

	FS1-DX (3F6)	FS1SE	FS1 (2RV, 3AU)
Overall length	1790 mm (70.5 in)	1900 mm (74.8 in)	1825 mm (71.9 in)
Overall width	775 mm (30.5 in)	825 mm (32.5 in)	750 mm (29.5 in)
Overall height	1040 mm (41.0 in)	1155 mm (45.5 in)	1015 mm (40.0 in)
Wheelbase	1160 mm (45.7 in)	1205 mm (47.4 in)	1160 mm (45.7 in)
Ground clearance – unladen	145 mm (5.7 in)	145 mm (5.7 in)	145 mm (5.7 in)
Weight	Dry–73 kg (161 lb)	Kerb–86 kg (190 lb)	Kerb–82 kg (181 lb) Dry–72 kg (159 lb)

Specifications relating to Chapter 1

Engine
Power output – restricted models .... 3 bhp @ 5500 rpm
Compression ratio:
    FS1E (394), FS1E-A, FS1E-DX (596), FS1E-DXA 7.1:1
    FS1/FS1M, FS1-DX/FS1M-DX (2GO) 6.9:1
    FS1 (2RV, 3AU), FS1-DX (3F6), FS1SE 6.6:1
Petrol/oil ratio – FS1E (394), FS1E-DX (596) only .... 20:1 See pages 11 and 47

Cylinder head
Gasket face maximum warpage .... 0.03 mm (0.0012 in)

Piston
Standard OD .... 40.0 mm (1.5748 in)
Piston/cylinder clearance .... 0.035 – 0.040 mm (0.0014 – 0.0016 in)
Service limit .... 0.1 mm (0.0039 in)

Cylinder barrel
Standard bore size .... 40.00 – 40.02 mm (1.5748 – 1.5756 in)
Maximum taper .... 0.05 mm (0.002 in)
Maximum ovality .... 0.01 mm (0.0004 in)

Piston rings
Type .... Keystone
Thickness .... 1.5 mm (0.06 in)
Width .... 1.8 mm (0.07 in)
End gap – installed .... 0.15 – 0.35 mm (0.006 – 0.014 in)
Ring/piston groove clearance .... 0.03 – 0.08 mm (0.001 – 0.003 in)

Crankshaft
Width across flywheels .... 37.90 – 37.95 mm (1.492 – 1.494 in)
Maximum runout .... 0.03 mm (0.001 in)
Big-end bearing deflection – at small-end .... 0.8 – 1.0 mm (0.03 – 0.04 in)
Service limit .... 2 mm (0.08 in)
Big-end side clearance:
    Standard – except FS1 (2RV, 3AU) .... 0.15 – 0.45 mm (0.006 – 0.018 in)
    Service limit – except FS1 (2RV, 3AU) .... 0.50 mm (0.020 in)
    FS1 (2RV, 3AU) .... 0.20 – 0.70 mm (0.008 – 0.028 in)

Clutch
Friction plate standard thickness .... 3.5 mm (0.14 in)
Service limit .... 3.2 mm (0.13 in)
Plain plate thickness .... 1.6 mm (0.06 in)
Plain plate maximum warpage .... 0.05 mm (0.002 in)
Clutch spring free length .... 34.0 mm (1.34 in)
Service limit:
    FS1 (2RV, 3AU) .... 31.0 mm (1.22 in)
    All other models .... 33.0 mm (1.30 in)

Kickstart
Friction clip resistance .... 0.8 – 1.2 kg (1.76 – 2.65 lb)

Gear ratios
Final drive – FS1 (2RV, 3AU), FS1-DX (3F6), FS1SE 3.000:1 (39/13)

Torque settings

Component	kgf m	lbf ft
Spark plug	1.5 – 2.5	11 – 18
Cylinder head nuts	0.8 – 1.0	6 – 7
Cylinder studs – FS1 (2RV, 3AU)	1.0	7
Crankcase and crankcase cover screws:		
FS1 (2RV, 3AU)	0.8	6
All other models	1.1 – 1.3	8 – 9.5
Generator rotor retaining nut:		
FS1 (2RV, 3AU)	4.5	32.5
FS1SE	3.0	22
All other models	5.0 – 7.0	36 – 50.5
Generator stator screws – FS1 (2RV, 3AU)	0.7	5
Primary drive gear retaining nut:		
FS1 (2RV, 3AU)	4.5	32.5
FS1SE	4.0	29
All other models	5.0 – 8.0	36 – 58

Component	kgf m	lbf ft
Disc valve cover screws – FS1 (2RV, 3AU)	0.8	6
Clutch spring screws:		
FS1 (2RV, 3AU), FS1SE	0.6	4
All other models	0.7 – 1.0	5 – 7
Clutch centre retaining nut:		
FS1 (2RV, 3AU)	4.5	32.5
FS1SE	4.0	29
All other models	5.0 – 7.0	36 – 50.5
Index arm shouldered bolt – FS1 (2RV, 3AU)	1.4	10
Bearing retainer plate screws – FS1 (2RV, 3AU)	0.8	6
Neutral indicator lamp switch – FS1 (2RV, 3AU)	2.0	14.5
Gearbox sprocket retaining nut:		
FS1E (394), FS1E-A, FS1E-DX (596), FS1E-DXA	6.4 – 10.0	46 – 72
FS1/FS1M (2GO), FS1-DX/FS1M-DX (2GO)	5.0 – 7.0	36 – 50.5
Kickstart crank pinch bolt – FS1 (2RV, 3AU)	1.2	9
Gearchange pedal pinch bolt – FS1 (2RV, 3AU), FS1SE	1.0	7
Transmission oil drain plug – FS1 (2RV, 3AU)	2.0	14.5
Engine mounting bolts:		
FS1 (2RV, 3AU)	2.4	17
FS1SE	2.5	18
All other models	2.2 – 3.0	16 – 22
Downtube mounting bolts:		
FS1 (2RV, 3AU)	2.4	17
FS1SE	2.5	18

Specifications relating to Chapter 2

Fuel tank capacity

FS1E (394), FS1E-A, FS1E-DX (596), FS1E-DXA	6.0 litres (1.32 gallons)
FS1/FS1M (2GO), FS1-DX/FS1M-DX (2GO)	6.5 litres (1.43 gallons)
FS1 (2RV, 3AU), FS1-DX (3F6), FS1SE	9.0 litres (1.98 gallons)

Carburettor

Make	Mikuni
Type	VM16SC
ID number:	
FS1E (394), FS1E-A	257E2
FS1E-DX (596), FS1E-DXA, FS1/FS1M (2GO),	
FS1-DX/FS1M-DX (2G0)	59600
FS1 (2RV, 3AU), FS1-DX (3F6), FS1SE	3F600
Needle jet – all models 1975 on	E4
Starter jet – FS1 (2RV, 3AU)	35

Air filter type

All models up to 1978	Dry, pleated paper
All models 1979 on	Dry polyurethane foam

Disc valve

Disc diameter	100.0 mm (3.94 in)
Disc standard thickness	3.0 mm (0.118 in)
Service limit	2.92 mm (0.115 in)
Disc maximum warpage	0.08 mm (0.003 in)
Valve opening duration	124°

Engine lubrication system

FS1E (394), FS1E-DX (596)	Petroil mixture, 20:1 ratio
All other models	Yamaha Autolube (pump fed total loss system)
Oil tank capacity:	
FS1E-A, FS1/FS1M (2GO), FS1E-DXA, FS1-DX/	
FS1M-DX (2GO)	1.15 litres (2.02 pints)
FS1 (2RV, 3AU), FS1-DX (3F6)	1.40 litres (2.46 pints)
FS1SE	1.30 litres (2.29 pints)
Recommended oil:	
Petroil lubrication	Good quality self-mixing two-stroke oil
Autolube oil tanks	Good quality two-stroke oil suitable for use in injection systems of air-cooled engines
Autolube pump minimum stroke	0.20 – 0.25 mm (0.0079 – 0.0098 in)

Gearbox lubrication

Capacity:
At oil change .... 600 – 650 cc (1.06 – 1.14 pints)
At engine rebuild .... 700 cc (1.23 pints)
Recommended oil .... Good quality SAE10W/30 SE engine oil

Torque settings – FS1 (2RV, 3AU) only

Component	kgf m	lbf ft
Oil pump mounting screws	0.8	6
Oil pipe banjo union bolt	1.0	7
Transmission oil drain plug	2.0	14.5
Exhaust pipe to cylinder ring nut	4.5	32.5
Exhaust pipe to silencer ring nut	4.0	29
Silencer mounting nut	2.0	14.5

Specifications relating to Chapter 3

Flywheel generator

Make:
FS1 (2RV, 3AU) .... Yamaha
All other models .... Mitsubishi
Type:
FS1E (394), FS1E-A, FS1/FS1M (2GO), FS1E-DX (596),
FS1E-DXA, FS1-DX/FS1M-DX (2GO) FOTO3171/4
FS1-DX (3F6) FOTO31
FS1SE FITI64
FS1 (2RV, 3AU) F355
Ignition source coil resistance – Black or Black/white to Earth
FS1SE .... 1.64 ohm ± 10% @ 20°C (68°F)
FS1 (2RV, 3AU) .... 1.4 – 2.0 ohm @ 20°C (68°F)
All other models .... 1.35 ohm ± 10% @ 20°C (68°F)
Condenser capacity:
FS1E (394), FS1E-A .... 0.22 microfarad ± 10%
FS1/FS1M (2GO), FS1-DX/FS1M-DX (2GO), FS1-DX
(3F6), FS1SE .... 0.25 microfarad ± 10%
All other models .... 0.30 microfarad ± 10%
Condenser minimum resistance:
FS1SE .... 5 M ohm
All other models .... 3 M ohm

Ignition HT coil

	FS1 (2RV, 3AU)	All other models
Make	Yamaha	Mitsubishi
Type	C481	F6T40177
Winding resistances @ 20°C (68°F):		
Primary – Black or Black/white to earth	1.40 – 1.80 ohm	1.02 ohm ± 10%
Secondary – HT lead to Earth	6.0 – 7.3 K ohm	6.0 K ohm ± 10%
Spark plug cap resistance @ 20°C (68°F)	5 K ohm	8.5 K ohm ± 20%

Spark plug – FS1 (2RV, 3AU) only

Make and type .... NGK BR7HS
Electrode gap .... 0.6 – 0.7 mm (0.024 – 0.028 in)

Torque settings

Component	kgf m	lbf ft
Spark plug	1.5 – 2.5	11 – 18
Generator rotor retaining nut:		
FS1 (2RV, 3AU)	4.5	32.5
FS1SE	3.0	22
All other models	5.0 – 7.0	36 – 50.5
Generator stator screws – FS1 (2RV, 3AU)	0.7	5

Specifications relating to Chapter 4

Front forks

Wheel travel:
Except FS1SE .... 85 mm (3.4 in)
FS1SE .... 110 mm (4.3 in)

Fork spring free length:
FS1E (394), FS1E-A, FS1/FS1M (2GO), FS1 (2RV,
3AU) – standard .... 153.0 mm (6.024 in)
FS1E (394), FS1E-A, FS1/FS1M (2GO), FS1 (2RV,
3AU) minimum .... 151.0 mm (5.945 in)
All DX models .... 150.0 mm (5.906 in)
FS1SE .... 305.2 mm (12.016 in)
Fork oil capacity – per leg:
FS1/FS1M (2GO) .... 137 cc (4.82 fl oz)
FS1 (2RV, 3AU) .... 160 cc (5.63 fl oz)
All DX models .... 134 cc (4.72 fl oz)
FS1SE .... 151 cc (5.32 fl oz)
Fork oil level* – FS1SE only .... 365.1 ± 4 mm (14.37 ± 0.16 in)
Recommended fork oil .... SAE10W/30 SE engine oil or SAE10 fork oil

Note that fork oil level is the distance between the top of the fork stanchion (inner tube) and the top of the oil, and is measured with a suitable dipstick when the fork top bolt, the spacer, the washer and the fork spring have been removed and the stanchion has been compressed fully into the fork lower leg

Rear suspension

Wheel travel:
FS1 (2RV, 3AU) .... 67 mm (2.6 in)
FS1SE .... 62 mm (2.4 in)
All other models .... 65 mm (2.5 in)
Spring free length:
FS1 (2RV, 3AU) .... 211.5 mm (8.327 in)
FS1SE .... 204.3 mm (8.043 in)
All other models .... 209.0 mm (8.228 in)
Swinging arm maximum free play – at fork ends .... 1.0 mm (0.039 in)

Torque settings

Component	kgf m	lbf ft
Handlebar clamp bolts – FS1 (2RV, 3AU)	1.2	9
Handlebar bottom clamp retaining nuts:		
FS1 (2RV, 3AU), FS1SE	3.0	22
All other models	2.3 – 3.7	16.5 – 27
Steering stem top bolt:		
FS1E (394), FS1E-A, FS1E-DX (596), FS1E-DXA	3.5 – 4.8	25 – 34.5
FS1/FS1M (2GO), FS1-DX/FS1M-DX (2GO), FS1-DX (3F6)	2.3 – 3.7	16.5 – 27
FS1 (2RV, 3AU)	3.0	22
FS1SE	4.0	29
Fork leg top bolt:		
FS1E (394), FS1E-A, FS1/FS1M (2GO)	2.3 – 3.7	16.5 – 27
FS1E-DX (596), FS1E-DXA	3.5 – 4.8	25 – 34.5
FS1-DX/FS1M-DX (2GO), FS1-DX (3F6)	3.5 – 4.5	25 – 32.5
FS1 (2RV, 3AU), FS1SE	3.0	22
Bottom yoke pinch bolts:		
FS1 (2RV, 3AU), FS1-DX (3F6)	2.5	18
FS1SE	3.0	22
All other models	1.4 – 2.4	10 – 17
Engine mounting bolts:		
FS1 (2RV, 3AU)	2.4	17
FS1SE	2.5	18
All other models	2.2 – 3.0	16 – 22
Downtube mounting bolts:		
FS1 (2RV, 3AU)	2.4	17
FS1SE	2.5	18
DX models	N/Av	N/Av
Seat mounting nuts and bolts:		
FS1 (2RV, 3AU)	1.0	7
FS1SE	N/Av	N/Av
All other models	0.8 – 1.3	6 – 9.5
Footrest mounting bolts:		
FS1 (2RV, 3AU)	2.4	17
All other models	1.8 – 2.9	13 – 21
Swinging arm pivot shaft nut:		
FS1E-DX (596), FS1E-DXA	2.0 – 3.0	14.5 – 22
FS1 (2RV, 3AU)	4.4	32
All other models	3.5 – 5.2	25 – 37.5
Rear suspension unit top mounting nut:		
FS1 (2RV, 3AU)	3.0	22
All other models	2.3 – 3.7	16.5 – 27

Component	kgf m	lbf ft
Rear suspension unit bottom mounting nut:		
FS1 (2RV, 3AU)	3.9	28
All other models	3.0 – 4.8	22 – 34.5

Specifications relating to Chapter 5

Wheels

Rim size – front:	
FS1E (394), FS1E-A, FS1/FS1M (2GO), FS1 (2RV, 3AU)	1.20 x 17
All DX models	1.40 x 17
FS1SE	1.40 x 19
Rim size – rear:	
Except FS1SE	1.40 x 17
FS1SE	1.60 x 16
Rim runout limit – radial:	
Except FS1SE	2.0 mm (0.08 in)
FS1SE	0.5 mm (0.02 in)
Rim runout limit – axial:	
Except FS1SE	2.0 mm (0.08 in)
FS1SE	1.0 mm (0.04 in)

Tyres

Size – front:	
FS1/FS1M (2GO), FS1 (2RV, 3AU), all DX models	2.50 x 17 - 4PR
FS1SE .:...	2.50 x 19 - 4PR
Size – rear	
Except FS1SE	2.50 x 17 - 4PR
FS1SE	3.00 x 16 - 4PR
Manufacturer's recommended minimum tread depth – front and rear, at centre of tread	1.0 mm (0.04 in)

Tyre pressures – FS1 (2RV, 3AU)

Front and rear, tyres cold .... 36 psi (2.5 kg/cm^2)

Note: maximum load (ie total weight of rider, passenger and any accessories or luggage) is 531 lb (241 kg)

Tyre pressures – all other models

	Front	Rear
Low speed, solo, maximum weight of rider, luggage and accessories not exceeding 198 lb (90 kg) – all models	23 psi (1.6 kg/cm^2)	28 psi (2.0 kg/cm^2)
High speed or with passenger, maximum load not exceeding 245 lb (111 kg) on the front tyre and 249 lb (113 kg) on the rear tyre:		
Except FS1SE	28 psi (2.0 kg/cm^2)	34 psi (2.4 kg/cm^2)
FS1SE	28 psi (2.0 kg/cm^2)	32 psi (2.25 kg/cm^2)

Brakes

	Front	Rear
Type:		
FS1E (394), FS1E-A, FS1/FS1M (2GO), FS1 (2RV, 3AU), FS1SE	Single leading shoe drum drum	Single leading shoe drum drum
All DX models	Hydraulic disc	Single leading shoe drum
Drum brake – front and rear, all models:		
Drum ID	110 mm (4.33 in)	
Service limit	111 mm (4.37 in)	
Brake shoe lining thickness	4.0 mm (0.16 in)	
Service limit	2.0 mm (0.08 in)	
Return spring free length	34.5 mm (1.36 in)	
Front disc brake – DX models only:		
Disc thickness	4.0 mm (0.16 in)	
Service limit	3.5 mm (0.14 in)	
Disc maximum runout	0.15 mm (0.006 in)	
Disc pad thickness – FS1E-DX (596), FS1E-DXA, FS1-DX/FS1M-DX (2GO):		
Moving (inner) pad	15.4 – 15.6 mm (0.60 – 0.61 in)	
Service limit	11.0 mm (0.43 in)	
Fixed (outer) pad	11.9 – 12.1 mm (0.47 – 0.48 in)	
Service limit	7.5 mm (0.30 in)	
Recommended hydraulic fluid	DOT 3 (US) or SAE J1703 (UK) brake fluid	

Torque settings

Component	kgf m	lbf ft
Front wheel spindle nut:		
FS1E-DX (596), FS1E-DXA	5.0 – 7.0	36 – 50.5
FS1-DX/FS1M-DX (2GO), FS1-DX (3F6)	5.0 – 6.0	36 – 43
FS1 (2RV, 3AU)	4.4	32
All other models	3.5 – 5.2	25 – 37.5
Front and rear brake operating arm pinch bolt –		
FS1 (2RV, 3AU)	0.9	6.5
Brake master cylinder/handlebar clamp bolts –		
FS1-DX/FS1M-DX (2GO), FS1-DX (3F6)	0.7 – 1.0	5 – 7
Brake hose union banjo bolts:		
FS1E-DX (596), FS1E-DXA	2.3 – 2.8	16.5 – 20
FS1-DX/FS1M-DX (2GO), FS1-DX (3F6)	2.3 – 3.7	16.5 – 27
Flexible brake hose upper/lower union – FS1E-DX (596),		
FS1E-DXA	1.5 – 2.0	11 – 14.5
Metal brake pipe union nuts:		
FS1E-DX (596), FS1E-DXA	1.3 – 1.8	9.5 – 13
FS1-DX/FS1M-DX (2GO)	1.3 – 2.1	9.5 – 15
Caliper bleed nipple	0.5 – 0.7	3.5 – 5
Fixed (outer) pad retaining screw – FS1E-DX (596),		
FS1E-DXA, FS1-DX/FS1M-DX (2GO)	0.15 – 0.25	1 – 1.5
Brake caliper to fork lower leg mounting bolts:		
FS1E-DX (596), FS1E-DXA	2.3 – 2.8	16.5 – 20
FS1-DX/FS1M-DX (2GO)	1.8 – 2.9	13 – 21
FS1-DX (3F6)	2.0	14.5
Brake caliper pivot retaining nut – FS1-DX (3F6)	2.5	18
Brake disc to wheel hub mounting bolts:		
FS1E-DX (596), FS1E-DXA	2.0 – 2.5	14.5 – 18
FS1-DX/FS1M-DX (2GO), FS1-DX (3F6)	1.8 – 2.9	13 – 21
Rear wheel spindle nut:		
FS1 (2RV, 3AU)	4.5	32.5
FS1E-DX (596), FS1E-DXA	5.0 – 6.0	36 – 43
All other models	5.0 – 7.0	36 – 50.5
Rear brake torque arm mountings – FS1 (2RV, 3AU)	1.8	13
Rear sprocket to hub mounting bolts or nuts:		
FS1 (2RV, 3AU)	2.4	17
FS1E-DX (596), FS1E-DXA	1.6 – 2.4	11.5 – 17
All other models	1.4 – 2.2	10 – 16

Specifications relating to Chapter 6

Generator

Output – FS1E (394), FS1E-A, FS1E-DX (596), FS1E-DXA:	
Charging coil – night only	0.5A min @ 2500 rpm (battery voltage 6.5V)
	4.0A max @ 8000 rpm (battery voltage 8.5V)
Lighting coil	6.3V min @ 2500 rpm (battery voltage 8.5V)
	8.7V max @ 8000 rpm (battery voltage 7V)
Output – FS1/FS1M (2GO), FS1-DX/FS1M-DX (2GO):	
Charging coil – day	0.5A min @ 2500 rpm
	4.0A max @ 8000 rpm
Charging coil – night	0.15 – 0.4A @ 8000 rpm
Lighting coil	6.0V min @ 2500 rpm
	8.5V max @ 8000 rpm
Output – FS1-DX (3F6):	
Charging coil – day	0.7A min @ 3000 rpm
	4.5A max @ 8000 rpm
Charging coil – night	0.2A min @ 3000 rpm
	3.0A max @ 8000 rpm
Lighting coil	6.0V min @ 2500 rpm
	8.0V max @ 8000 rpm
Output – FS1SE:	
Charging coil – day	1.3A min @ 3000 rpm
	2.0A max @ 8000 rpm
Charging coil – night	0.8A min @ 3000 rpm
	2.0A max @ 8000 rpm
Lighting coil	5.8V min @ 2500 rpm
	8.5V max @ 8000 rpm

Output – FS1 (2RV, 3AU):
    Charging coil – day .... 0.7A min @ 3000 rpm
                                      4.7A max @ 8000 rpm
    Charging coil – night .... 0.35A min @ 3000 rpm
                                        2.5A max @ 8000 rpm
    Lighting coil .... 5.8V min @ 3000 rpm
                                        8.2V max @ 8000 rpm

Charging/lighting coil resistance values – at 20°C (68°F):
    FS1E (394), FS1E-A, FS1E-DX (596), FS1E-DXA, FS1/FS1M
    (2GO), FS1-DX/FS1M-DX (2GO):
        Charging – Green to Earth .... 0.30 ohm ± 10%
        Charging/lighting – Yellow to Earth .... 0.30 ohm ± 10%
    FS1 (2RV, 3AU):
        Charging – Green/red to Earth (Black) .... 0.30 – 0.50 ohm
        Charging/lighting – Yellow to Earth (Black) .... 0.40 – 0.60 ohm
    FS1-DX (3F6):
        Charging – Green/red to Earth .... 0.46 ohm ± 10%
        Charging/lighting – Yellow to Earth .... 0.26 ohm ± 10%
    FS1SE:
        Charging – White to Earth (Black) .... 0.36 ohm
        Lighting – Yellow/red to Earth (Black) .... 0.26 ohm ± 10%

Battery
    Make .... Yuasa, GS, or FB
    Type 6N4A-4D
    Capacity 6V 4AH

Voltage regulator – FS1SE only
    Make .... Stanley
    Type SU208Y
    No-load regulated voltage .... 7.5 ± 0.3V

Bulbs
    Headlamp:
        FS1SE 6V, 25/25W
        All other models 7V, 18/18W
    Stop/tail lamp .... 6V, 21/5W (some models may use 6V, 17/5.3W bulbs – check
                                        bulb holder marking to ensure correct bulb is fitted)

    Flashing indicator lamps:
        FS1E (394), FS1E-A, FS1E-DX (596), FS1E-DXA 6V, 8W
        FS1 (2RV, 3AU) 6V, 21W
        All other models 6V, 10W
    Speedometer illuminating lamp* 6V, 3W
    Flashing indicator and main beam warning lamps –
    FS1 (2RV, 3AU) 6V, 3W

Note: FS1 (2RV, 3AU) – Speedometer-mounted bulbs are of capless type; pull out of bulb holder to remove and press in to refit – take care not to damage fine wire tails

---

1 Introduction

The first six Chapters of this Manual relate to the standard FS1E moped. This Chapter describes the differences found when working on the later models. To assist owners in identifying their machines correctly, this Section describes the modifications as they were introduced.

The standard FS1E model continued unchanged apart from minor styling modifications such as higher, braced handlebars, the letters 'FS1E' instead of 'SS' on the sidepanels, and purple instead of candy gold paint.

1975

FS1E (Yamaha model code number 394) – continued unchanged apart from a colour change to brown from August onwards. Flashing indicators were not fitted as standard but were available in kit form from the importers.

FS1E-DX (Yamaha model code number 596) – introduced in May. Fitted with stronger front forks and an hydraulically operated disc brake; this necessitated fitting a larger-section front tyre, a restyled front mudguard and a separate speedometer drive gearbox on the hub left-hand side. The ignition switch was moved to the fork top yoke, next to the speedometer, flatter unbraced handlebars were fitted, flashing indicator lamps were fitted as standard, as was a helmet lock and a mirror. The choke control was moved from the handlebar

to the carburettor top. The new model was available in bright yellow only.

1976

Both FS1E and FS1E-DX continued unchanged apart from minor styling alterations.

1977

The FS1E and FS1E-DX continued unchanged until they were discontinued later in the year.

In April the FS1E-A and FS1E-DXA models were introduced. These were similar to the previous models, but were fitted with Yamaha Autolube engine lubrication, in which oil carried in a separate tank behind the right-hand sidepanel is fed to a pump mounted beneath the carburettor. The pump is driven by gears from the crankshaft and supplies oil to an injection nozzle in the inlet tract. A cable linked to the throttle ensures that the pump's output varies according to the engine's needs. The system is more efficient than the previous petroil lubrication and ensures that decoking intervals can be extended, apart from the elimination of the need to mix petrol and oil whenever the tank is refilled. The FS1E-A and FS1E-DXA were discontinued later in the year.

As a result of new legislation, mopeds were re-defined as lightweight motorcycles of no more than 50 cc engine capacity with a designed maximum speed of no more than 30 mph. Yamaha's answer was to introduce two new, restricted models

1a Reservoir and master cylinder –
FS1E-DX (596) and FS1E-DXA. Note
early type stop lamp switch

1b Front brake caliper – FS1E-DX
(596), FS1E-DXA and
FS1-DX/FS1M-DX (2GO)

1c Ignition switch moved to top yoke
– FS1E-DX (596)

1d Autolube models – oil is contained
in separate tank behind right-hand
side panel ...

1e ... and is injected via nozzle into
inlet tract

1f Plate on steering head gives details
of all restricted models ...

1g ... which are fitted with footrests
instead of pedals

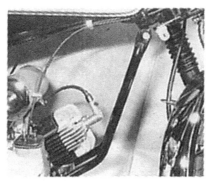

1h Cosmetic downtubes fitted as
standard on some models

1i Remove rubber plug to inspect
brake shoe friction material – 2GO,
3F6, 2RV and 3AU models

1j Note new type of rear stop lamp
switch ...

1k ... fitted in swinging arm – 2GO
models

in August 1977. The new standard model was the FS1M, its disc-braked counterpart being the FS1M-DX. The new models were fitted with footrests instead of pedals, but apart from their restricted power outputs and slightly different styling, were much the same as their predecessors. Flashing indicator lamps of a slightly different pattern were fitted as standard. The DX model was fitted with a tamperproof master cylinder reservoir cover and two bolt-on tubes running from beneath the steering head to lugs on the crankcase front. This is a purely cosmetic feature fulfilling no functional purpose. The tubes were previously available from proprietary manufacturers or as optional extras from the importers, and so may be encountered on earlier models. The manufacturer's model code for both new models was 2GO.

1978

Both FS1M and FS1M-DX continued unchanged, apart from differing paint schemes. Also referred to as the FS1 and FS1-DX, respectively.

1979

The FS1M (FS1) continued unchanged until October, when it was phased out.

The FS1M-DX was replaced by the FS1-DX (Yamaha model code number 3F6; engine/frame numbers begin at 3F6-000101). Although substantially similar to the previous model, the new version employed a modified brake master cylinder and a new type of caliper mounted behind the fork leg, a modified cylinder head and barrel (recognisable by the larger cooling fins), a larger fuel tank, a different seat which was fitted with a plastic rear cover (attached by four nuts), and various other styling alterations. In addition, a polyurethane foam air filter in a black plastic casing replaced the previous filter assembly, and a self-adjusting junction box was fitted between the upper throttle cable and the lower throttle and oil pump cables.

1980

The FS1-DX continued unchanged (engine/frame numbers begin at 3F6-100101).

1981

The FS1-DX continued unchanged. A new model, the FS1SE (Yamaha model code number 5A1), was introduced. Basically sharing the engine/gearbox unit and most other parts of the FS1-DX, the FS1SE was finished in the increasingly popular 'factory custom' style. It had high handlebars, extended front forks, a teardrop tank, stepped seat and other styling modifications. The wheel sizes were changed, drum brakes fitted front and rear, black plastic flashing indicator lamps employed, and the engine oil tank was moved to the rear of the seat.

1982

Both the FS1-DX and FS1SE continued unchanged.

1983

Both the FS1-DX and FS1SE remained unaltered until they were discontinued, the FS1-DX in April and the FS1SE in October.

1987

The FS1 was reintroduced in March. The model code for the 1987 model is 2RV (engine/frame numbers begin at 3F6-105101 on). The new model can be distinguished easily by its square-bodied flashing indicator lamps; apart from the FS1E/FS1/FS1M-pattern front forks and front drum brake, it is otherwise almost identical to the FS1-DX (3F6) model.

1988

The FS1 continued unchanged, but its model code is now 3AU. Engine/frame numbers begin with 3F6-112101.

1989 on

With the exception of fuel tank and sidepanel graphics the FS1 continues unchanged. Its model code is slightly altered to 3AU1, but reference should be made to the 3AU information throughout this Chapter. Engine and frame numbers begin at 3F6-116101. The model can be distinguished from its predecessor by the letters FS1 instead of 50 on the sidepanels, and by the fitting of either red or yellow fork gaiters.

2 Routine maintenance – schedule – all 1977 on models

Note that revised service schedules have been introduced for all later models, from the FS1/FS1M (2GO) and FS1-DX/FS1M-DX (2GO) onwards. Owners should proceed as follows, using either the mileage or time interval, whichever occurs first.

Daily, pre-ride check:
Check the transmission oil level (gearbox)
Check the engine oil level (oil tank)
Check the fuel level
Check the wheels and tyres, especially tyre pressures and tread wear
Check the brakes
Check all controls are properly adjusted and working correctly
Check that the final drive chain is well lubricated and correctly adjusted
Check the tightness of all nuts, bolts and other fasteners
Check that the speedometer, horn and all lights are working correctly.

Monthly or every 300 miles (500 km)
Check carefully all items listed under the daily check
Clean, lubricate and adjust the final drive chain
Check the spark plug condition
Check the battery electrolyte level
Check the brake hydraulic fluid level – DX models only
If the machine is used in wet or dusty conditions, clean the air filter element

Six monthly or every 2000 miles (3000 km):
Change the transmission oil
Clean and adjust the spark plug
Clean the air filter element
Clean the silencer baffle tube
Clean and check the contact breaker points. Oil the cam lubricating felt
Check the engine idle speed and carburettor settings, adjust the throttle cable and check the oil pump minimum stroke setting, and either the cable adjustment (2GO models) or the cable operation (later models)
Grease the throttle twistgrip and lubricate all control cables and pivots
Check, clean and adjust or renew (as necessary) the brakes. Check the hydraulic fluid level and ensure that the caliper is free to slide on its mounting pins (2GO) or to pivot on its mounting (3F6); renew the pads if they are worn to the limits or beyond. Grease the (drum) brake camshafts and pedal or lever pivots
Check the front and rear suspension
Check the final drive chain
Check the wheels and tyres
Check the battery, lights and horn
Check and adjust the clutch
Check the tightness of all nuts, bolts and other fasteners

Annually or every 4000 miles (6000 km)
Repeat all tasks given under previous checks, then carry out the following:
Renew the spark plug
Renew the air filter element, if necessary
Decoke the cylinder head, barrel, piston and exhaust.

Renew the piston rings (if required) and all gaskets and seals disturbed

Clean the fuel tap (and tank, if necessary) and carburettor

Check the contact breaker points and renew them if necessary. Check and adjust the ignition timing

Grease the steering head bearings and check adjustment

Change the front fork oil

Remove the wheels, check the brake components (clean and renew or refit, as necessary, applying a smear of high melting-point grease as described to any bearing surfaces), check and grease the wheel bearings and speedometer drive

Remove, clean and grease the swinging arm pivot shaft and centre stand pivot shaft

Renew the hydraulic fluid – DX models only

Additional items

All DX models – at regular intervals (usually every two to four years) the brake hose(s) and the master cylinder and caliper seals must be renewed, regardless of their apparent condition. This is recommended as a safety measure, to prevent the risk of sudden brake failure due to the ageing of these components.

---

3 Engine – modifications

Primary drive gears – renewal

1   When obtaining new primary drive parts note that the two components are matched to give a prescribed amount of

backlash. Ensure that the match marks etched on the inner face of each correspond to avoid excessive or insufficient clearance. The marks, in the form of the letters 'B', 'C', or 'D' must be the same on both components.

Clutch

2   With reference to Fig. 1.3, note that the kickstart driven pinion (item 23) is integral with the clutch outer drum (item 1) on all models from 1977 (2GO). Note also that the clutch centre nut lock washer (item 12) is of the belleville type on all later models; this washer must be fitted with its convex face outwards.

Gearbox

3   With reference to Fig. 7.1, note the various differences in the gearbox components of the 1979-on models. The most significant change is the method of securing the gearbox sprocket, which is now retained by a circlip.

Selector drum

4   With reference to Fig. 1.8, note that the special locating pin and circlip (items 10 and 11) are replaced by a roll pin on all models from 1979 on, ie FS1-DX (3F6), FS1 (2RV, 3AU) and FS1SE.

---

4 Air filter – cleaning – FS1-DX (3F6), FS1 (2RV, 3AU) and FS1SE

---

1   The later type of air filter incorporates a polyurethane foam element mounted on top of the engine/gearbox unit in a black

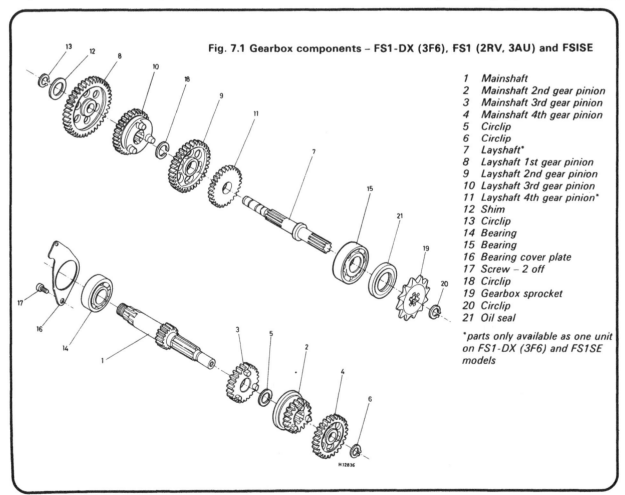

Fig. 7.1 Gearbox components – FS1-DX (3F6), FS1 (2RV, 3AU) and FSISE

1   Mainshaft
2   Mainshaft 2nd gear pinion
3   Mainshaft 3rd gear pinion
4   Mainshaft 4th gear pinion
5   Circlip
6   Circlip
7   Layshaft*
8   Layshaft 1st gear pinion
9   Layshaft 2nd gear pinion
10   Layshaft 3rd gear pinion
11   Layshaft 4th gear pinion*
12   Shim
13   Circlip
14   Bearing
15   Bearing
16   Bearing cover plate
17   Screw – 2 off
18   Circlip
19   Gearbox sprocket
20   Circlip
21   Oil seal

*parts only available as one unit on FS1-DX (3F6) and FS1SE models

H.12836

plastic casing. The filter must be cleaned every 2000 miles (3000 km) or sooner if the machine is used in wet or dusty conditions.

2 Remove the single screw securing the filter cover and withdraw the cover, disengaging it from the filter/crankcase cover rubber hose and the HT lead and fuel feed pipe. Remove the foam element. If it is split or torn, or too choked to be reusable, it must be renewed. To clean the element, soak it in a high flash-point solvent such as white spirit, ensuring that suitable precautions are taken to prevent the risk of fire. Dry the element by squeezing it gently and allowing surplus solvent to evaporate; do not wring it out or it may be damaged. Alternatively, compressed air may be used to blow clear the element.

3 Check that the element supporting wire mesh is correctly refitted, place the element on top and refit the cover, ensuring that it engages correctly with the rubber hose and that the securing screw is not overtightened. A light smear of grease to the cover sealing surface will help prevent the entry of unfiltered air; check also that the coil spring which seals the cover/hose joint is correctly refitted.

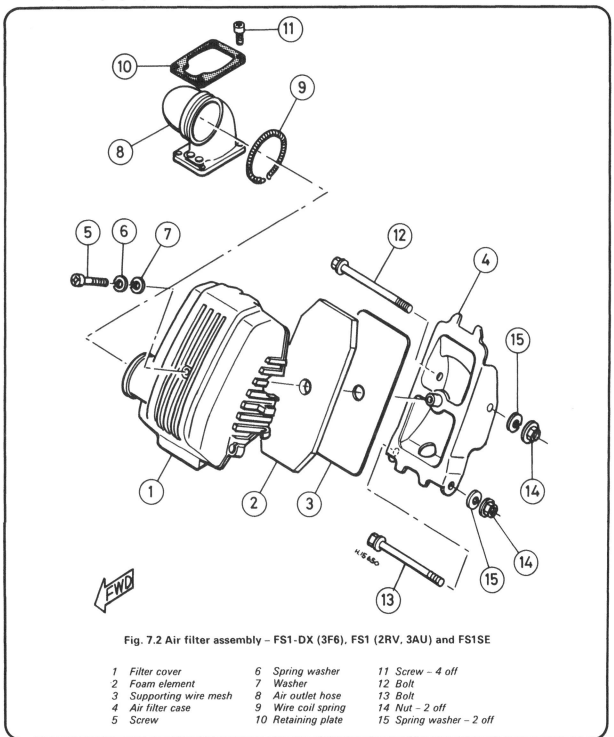

Fig. 7.2 Air filter assembly – FS1-DX (3F6), FS1 (2RV, 3AU) and FS1SE

1 Filter cover	6 Spring washer	11 Screw – 4 off
2 Foam element	7 Washer	12 Bolt
3 Supporting wire mesh	8 Air outlet hose	13 Bolt
4 Air filter case	9 Wire coil spring	14 Nut – 2 off
5 Screw	10 Retaining plate	15 Spring washer – 2 off

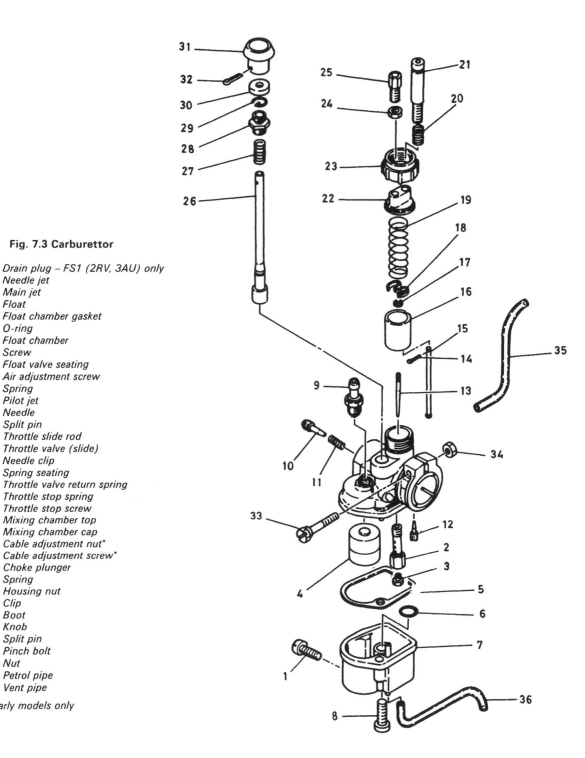

Fig. 7.3 Carburettor

1   Drain plug – FS1 (2RV, 3AU) only
2   Needle jet
3   Main jet
4   Float
5   Float chamber gasket
6   O-ring
7   Float chamber
8   Screw
9   Float valve seating
10 Air adjustment screw
11 Spring
12 Pilot jet
13 Needle
14 Split pin
15 Throttle slide rod
16 Throttle valve (slide)
17 Needle clip
18 Spring seating
19 Throttle valve return spring
20 Throttle stop spring
21 Throttle stop screw
22 Mixing chamber top
23 Mixing chamber cap
24 Cable adjustment nut*
25 Cable adjustment screw*
26 Choke plunger
27 Spring
28 Housing nut
29 Clip
30 Boot
31 Knob
32 Split pin
33 Pinch bolt
34 Nut
35 Petrol pipe
36 Vent pipe

* early models only

5 Exhaust system – cleaning – FS1/FS1M (2GO, 2RV, 3AU) and FS1-DX/FS1M-DX (2GO, 3F6)

While the exhaust system of the later models is also cleaned exactly as described in Chapter 2, owners shoud note that a baffle plate (approximately) 400 – 460 mm (16 – 18 in) from the silencer rear end requires particular attention. The plate has a hole 13 mm (0.5 in) in diameter drilled in it which can cause a considerable reduction in top speed if it is allowed to become even partially blocked by carbon build-up. Owners are advised to use a rod of suitable length and diameter to ensure that the hole is clear whenever the baffle tube is removed for cleaning.

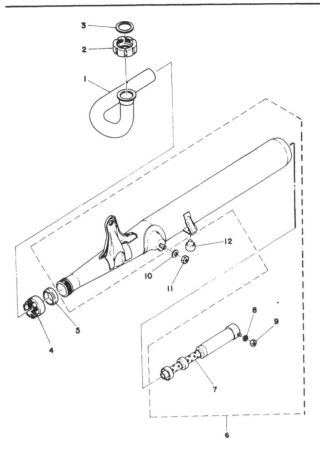

Fig. 7.4 Exhaust system

1	Exhaust pipe	7	Baffle tube
2	Ring nut	8	Spring washer
3	Gasket	9	Nut
4	Silencer nut	10	Spring washer
5	Gasket	11	Nut
6	Silencer assembly	12	Centre stand stop

6 Autolube pump – maintenance and adjustment

Pump removal and refitting

1   The pump is mounted on the crankcase left-hand cover by two screws. Remove the carburettor cover and its gasket to gain access to the pump. Plug both oil feed pipes as soon as they are disconnected to prevent the loss of oil or the entry of air, disengage the pump cable from the pump pulley, remove the two screws and withdraw the pump, noting the shim on the drive shaft end.

2   Refitting is a reversal of the above procedure. Ensure that the shim is refitted on the drive shaft end and that a new gasket is used. Tighten the mounting screws securely and connect the pump cable and oil feed pipes, then bleed the pump and check its stroke setting and cable adjustment as described below.

3   Note that the Autolube pump is a sealed unit; if any fault occurs, or if the pump's efficiency is suspect, it must be renewed.

4   The drive mechanism can be reached only after the crankcase left-hand cover has been removed, but should not be disturbed unless necessary; wear is unlikely until a very high mileage has been covered.

6.1a Oil pump is mounted on crankcase right-hand cover

6.1b Do not omit shim on pump worm shaft end

6.4a Pump drive can be dismantled once crankcase cover is withdrawn

6.4b Tighten oil feed pipe unions securely – check routing

Bleeding the system

5   This must be carried out whenever the Autolube tank has been allowed to run dry, or any part of the system has been disconnected.

6   To bleed the oil tank/pump feed pipe and the pump, firstly fill the tank with 2-stroke oil, then remove the carburettor cover and the bleed screw. Let the oil run out until all the air bubbles disappear. When there are no more bubbles tighten the bleed screw and install the pump cover. Check the bleed screw gasket, and if damaged, replace it with a new one.

7   The pump delivery pipe can be bled by starting the engine and pulling the pump wire all the way out to set the pump stroke to a maximum, then keeping the engine running at about 2000 rpm for two minutes or so to clear all the air out of the system.

Checking and adjusting the pump minimum stroke

8   While the pump minimum stroke setting must be checked every 2000 miles (3000 km), the actual adjustment is unlikely to be required very often.

9   To check the setting, remove the carburettor cover, start the engine and allow it to idle. On looking at the rear of the pump, it will be seen that the pump adjustment plate moves in and out. When the plate is out to its fullest extent, stop the engine and measure the gap between the plate and the raised boss of the pump pulley using feeler gauges. Do not force the feeler gauge into the gap – it should be a light sliding fit. Make a note of the reading, then repeat the procedure several times. The largest gap is indicative that the pump is at its minimum stroke position. If the pump is set up correctly, the gap found should be 0.20 – 0.25 mm (0.008 – 0.010 in).

10   If the pump setting is found to be incorrect, remove the adjustment plate retaining nut and withdraw the spring washer and the adjustment plate. The pump stroke is set by adding or removing shims from behind the adjustment plate, these shims being available from Yamaha dealers in thicknesses of 0.3, 0.5 and 1.0 mm (0.0118, 0.0197 and 0.0394 in). When the shim thickness is correct, refit the adjustment plate, the spring washer, and the retaining nut, then start the engine and recheck the minimum stroke setting. If necessary, repeat the procedure to ensure that the setting is correct.

Oil pump cable adjustment – FS1E-A, FS1E-DXA, FS1/FS1M (2GO) and FS1-DX/FS1M-DX (2GO)

11   To ensure that the oil pump remains accurately synchronised with the carburettor, the pump cable adjustment must be checked every 2000 miles (3000 km). This work must only be carried out after the carburettor settings have been checked, and reset if necessary. Check that the throttle cable has 0.5 – 1.0 mm (0.02 – 0.04 in) of free play. Remove the carburettor cover.

12   Slowly open the throttle until the top edge of the indentation drilled in the throttle slide aligns with the top of the carburettor bore, as shown in the accompanying photograph. At this point the pump plunger pin should align exactly with the third of the pulley reference marks (see accompanying photograph). If adjustment is required, slacken the adjuster locknut and rotate the cable adjuster until the pin and reference mark are aligned. Tighten the locknut securely, open fully the throttle once or twice to settle the cable, then check that the setting has not altered.

13   When the setting is correct, lubricate the pump cable inner wire and the moving parts of the pump, then refit the carburettor cover and gasket.

Checking operation of oil pump cable – FS1-DX (3F6), FS1 (2RV, 3AU) and FS1SE

14   These models are fitted with an oil pump whose setting is controlled by a cable connected via a junction box to the throttle cable, this junction box being of a new type which automatically compensates for wear in the cables and ensures that the oil pump is synchronised accurately with the carburettor at all times. The junction box is discussed separately in Section 7 of this Chapter. All that can be done at the oil pump is to check that the cable is operating correctly, a task which must be carried out after the adjustment of the throttle cable has been checked, and reset if necessary.

15   When the throttle cable has been checked and if necessary adjusted, and the carburettor cover has been removed, start the engine and allow it to idle until the pump adjustment plate moves out to its fullest extent, ie to the minimum stroke position, as described above, so that the pulley is free to rotate throughout its full movement. Stop the engine and allow the twistgrip to rotate forwards to the fully closed position. Slowly rotate the twistgrip to open the throttle and watch the pump pulley closely. The pulley should start to rotate just as the

6.6 Bleed screw is provided to remove air from pump and tank/pump feed pipe

6.12a Open twistgrip until throttle slide mark is aligned as shown ...

6.12b ... when pulley guide pin should align with pulley mark shown

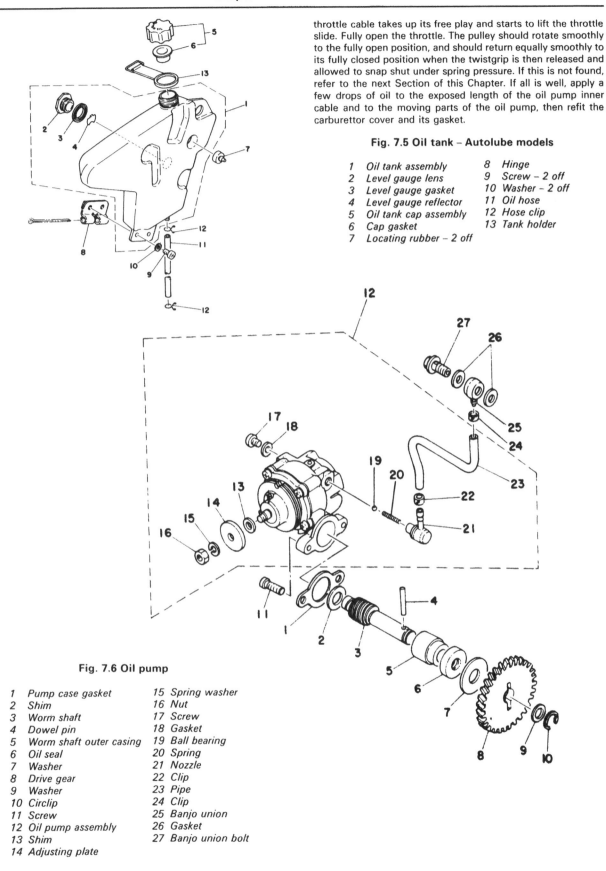

throttle cable takes up its free play and starts to lift the throttle slide. Fully open the throttle. The pulley should rotate smoothly to the fully open position, and should return equally smoothly to its fully closed position when the twistgrip is then released and allowed to snap shut under spring pressure. If this is not found, refer to the next Section of this Chapter. If all is well, apply a few drops of oil to the exposed length of the oil pump inner cable and to the moving parts of the oil pump, then refit the carburettor cover and its gasket.

Fig. 7.5 Oil tank – Autolube models

1	Oil tank assembly	8	Hinge
2	Level gauge lens	9	Screw – 2 off
3	Level gauge gasket	10	Washer – 2 off
4	Level gauge reflector	11	Oil hose
5	Oil tank cap assembly	12	Hose clip
6	Cap gasket	13	Tank holder
7	Locating rubber – 2 off		

Fig. 7.6 Oil pump

1	Pump case gasket	15	Spring washer
2	Shim	16	Nut
3	Worm shaft	17	Screw
4	Dowel pin	18	Gasket
5	Worm shaft outer casing	19	Ball bearing
6	Oil seal	20	Spring
7	Washer	21	Nozzle
8	Drive gear	22	Clip
9	Washer	23	Pipe
10	Circlip	24	Clip
11	Screw	25	Banjo union
12	Oil pump assembly	26	Gasket
13	Shim	27	Banjo union bolt
14	Adjusting plate		

7 Throttle/oil pump cable junction box – description and maintenance – FS1-DX (3F6), FS1 (2RV, 3AU) and FS1SE

1   To ensure that the synchronisation of the carburettor and oil pump remains exactly set without the need for constant checking and adjustment, Yamaha introduced on the 1979 model FS1-DX (3F6), a junction box which compensates automatically for wear in either the junction box/carburettor throttle cable or the junction box/oil pump cable.

2   The principle by which this operates is shown in figure 7.8, part 'A' and 'B'. In a normal junction box (part 'A') the throttle cable lifts a one-piece slider to which the carburettor cable and oil pump cable are attached. If all cables are correctly adjusted, and checked regularly to ensure this, the junction box will work very well. If, on the other hand, excessive free play is allowed to creep in through infrequent or incorrect adjustment, the oil pump will no longer be synchronised correctly with the carburettor, and will deliver too much or too little oil, depending on which cable is incorrectly set. Yamaha's new design, shown in basic form in figure 'B', introduces a rotating link instead of the slider. As the twistgrip is rotated, the link also rotates to take up the excess free play in whichever cable is worn.

3   A detailed description of the junction box's method of operation is as follows. Refer to figure 7.7 for identification of the parts concerned. The twistgrip/junction box inner cable is connected to the slider via the rotor guide, its nipple being retained in position by the set pin. As the slider moves up the junction box, the rotor turns in the rotor guide taking up the free play in the cables. A small compression spring set in the slider acts on the rotor guide, permitting the rotor to turn until its pressure is overcome by the greater resistance of the throttle slide and oil pump pulley return springs when the cable free play is eliminated, so that a raised locking tab on the rear of the slider comes into contact with serrations on the rotor periphery and locks it solidly. The junction box then opens the throttle slide and oil pump together in the normal way.

4   Referring to figure 7.9, in part A the oil pump cable is shown as having excessive free play. In part B the rotor is turning to take up this free play as the slider moves up the junction box, and in part C the free play has been eliminated and the slider lock has come up to bear on the rotor so that it cannot move.

5   It will be evident from the above description that the whole assembly relies on each cable and component being free from dirt or corrosion if it is to operate correctly. If the checks described in Section 6 of this Chapter reveal any sign that the assembly is not functioning correctly, remedial action must be taken at once. While the work can be carried out on the machine, it is very awkward and there is a high risk of dirt getting in, or of one of the smaller components being lost; it is therefore recommended that the complete throttle/oil pump cable assembly be removed before work starts. Disconnect the

throttle cable from the carburettor and the oil pump cable from the oil pump, then free the upper end of the cable from the twistgrip by slackening and removing the two twistgrip clamping screws. Remove the screw clamping the junction box to the frame and withdraw the complete assembly from the machine.

6   Slide back the rubber sleeves from the junction box, then slacken and remove the two screws which secure the box cover, and withdraw the cover. Turn the set pin through $\frac{1}{2}$ turn and withdraw it so that the twistgrip cable and nipple can be disengaged from the rotor guide, then very carefully pick out the rotor guide, ensuring that the small spring does not fly out of the slider. Pick out the rotor and disengage the end nipples of the two lower cables from it. The slider can then be picked out.

7   Carefully inspect the cables and junction box components for signs of wear or damage, renewing any individual component which is no longer serviceable; all are available through Yamaha dealers. Carefully clean off all traces of dirt and corrosion. Thoroughly lubricate the cables as described in

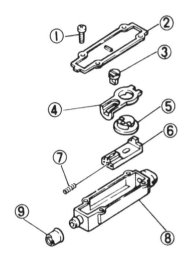

Fig. 7.7 Throttle/oil pump cable junction box – FS1-DX (3F6), FS1 (2RV, 3AU) and FS1SE

1	Screw – 2 off	6 Slider
2	Cover	7 Spring
3	Set pin	8 Junction box
4	Rotor guide	9 Rubber boot
5	Rotor	

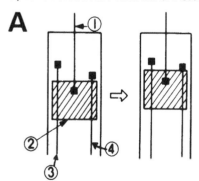

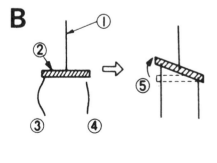

Fig. 7.8 Throttle/oil pump cable junction box operation – FS1-DX (3F6), FS1 (2RV, 3AU) and FS1SE

1	Throttle cable	3	Carburettor cable	5 Self adjustment method
2	Slider	4	Oil pump cable	

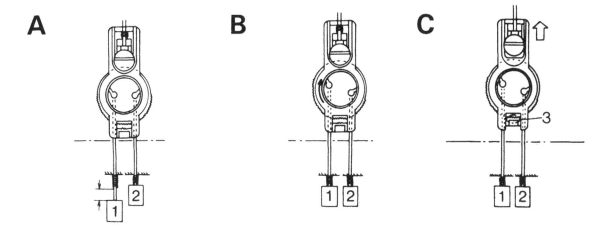

Fig. 7.9 Throttle/oil pump cable junction box self adjustment method – FS1-DX (3F6), FS1 (2RV, 3AU) and FS1SE

1   *Oil pump cable*   2   *Carburettor cable*   3   *Slider block*

Routine maintenance then reassemble the cables and junction box by reversing the dismantling procedure. Do not apply any oil or grease to the junction box components, but use instead a synthetic aerosol lubricant such as WD40, or one of the proprietary silicone based lubricants suitable for use with nylon components. Connect the cables to the carburettor and oil pump.

8   Pass the upper end of the throttle cable up to the handlebar ensuring that it is correctly routed with no sharp turns or kinks, then connect it to the twistgrip again and use the adjuster below the twistgrip to set the required amount of free play in the cable.

9   Open and close fully the twistgrip several times to settle the cables. Check the adjustments again and reset them if necessary, but if all is in order, tighten the adjuster locknuts and replace the protecting rubber covers. Make a final check that the oil pump cable is rotating the pump pulley correctly, as described in Section 6 of this Chapter, before refitting the carburettor cover.

8   Ignition HT coil – diode location

1   On some models (see relevant wiring diagram), a diode is fitted in the LT lead (Black or Black/white wire) between the generator and the ignition HT coil. On some machines it is a clearly recognisable tubular component, while on others it may be bound up in the main wiring loom; in all cases the diode will be found attached to the main wiring loom somewhere between the generator lead block connector and the HT coil – it will therefore be necessary to remove the battery and its carrier to reach the diode.

2   If the ignition breaks down at any time, and the cause is traced to a fault in the wire from the generator to the coil the diode should be suspected. To test the diode, check that current can flow in one direction only, if resistance is encountered in both directions, or in none at all, the diode must be renewed.

3   The diode is listed separately only for FS1-DX (3F6) 1979 and FS1SE models; for all other machines seek the advice of a competent Yamaha agent. It may be possible to use the same diode on other models.

9   Contact breaker points – gap setting

1   On FS1/FS1M (2GO), FS1-DX/FS1M-DX (2GO, 3F6) and

FS1SE models, two methods may be used. Either the traditional manner using a feeler gauge as explained in Chapter 3.4 or by setting the angle of dwell using the static pointer on the generator stator and the two stamped marks on the flywheel rotor. The latter method is much more accurate and is to be preferred.

2   First rotate the flywheel rotor in the direction indicated by the stamp-marked arrow until the first of the marks is aligned with the pointer. At this instant the contact breaker points will begin to open. Continue to rotate the flyweel rotor and when the next mark is aligned with the pointer the contact breaker points should just have closed.

3   To check this, a points checker, multimeter, or a simple battery and bulb set up may be used. Connect one end to the black lead which runs to the contact breaker points and the other to an earth point on the machine. Again, rotate the flywheel rotor. When the first mark is aligned with the pointer the reading on the multimeter should immediately cease (or the bulb should dim). Continue to rotate the flywheel until the next mark is aligned. The multimeter should then give a reading (or the bulb should immediately brighten). If this happens the points gap is correct. If it does not, the following action should

9.2a First mark passing fixed pointer – points should be opening

9.2b Second mark passing fixed pointer – points should just be closing

be taken to correct it. If a reading is given (or the bulb stays bright) after the first mark but ceases (or dims) before the second, make the gap larger and vice-versa. When the adjustment has been made, tighten everything and then re-check.

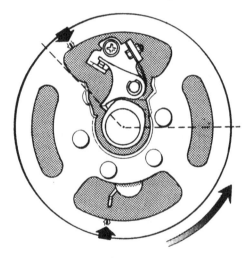

Fig. 7.10 Two marks and static pointer used to set dwell angle – FS1/FS1M (2GO), FS1-DX/FS1M-DX (2GO, 3F6) and FS1SE

10 Front forks and steering head – general

Front forks

1 The forks fitted to the FS1E-A and FS1/FS1M (2GO) models are similar in design and construction to those described in Chapter 4 of this Manual; apart from noting the different fork oil quantities, and fork spring free lengths given in the Specifications Section of this Chapter, no further information is necessary.

2 The other models are fitted with forks of differing types, although in general these differences are minor. Reference to the appropriate drawing accompanying the text before work commences will clarify any problems likely to occur during dismantling and reassembly. Note the different fork oil quantities and fork spring free lengths given in the Specifications Section of this Chapter.

3 The FS1E-DX (596), FS1E-DXA, FS1 (2RV, 3AU) and

FS1SE models are all fitted with forks that are essentially similar, the fork springs being external on the first three and internal (ie inside the stanchions) on the last. The major difference between both of these and the standard forks described in Chapter 4 is the method of retaining the oil seal. This is no longer held in the plated, screwed collar, but is pressed into a recess in the fork lower leg, retained by a circlip and a plain washer. Note that DX models are fitted with a separate tapered restrictor.

4 The circlip and washer must be withdrawn before the stanchion (inner tube) is pulled sharply out of the lower leg complete with the seal and bushes. Reassembly is straightforward, but the oil seal must be tapped carefully into the fork lower leg after the stanchion and bushes have been refitted; the seal must be tapped in only far enough for the plain washer to be refitted and for the circlip to be seated in its groove.

5 The FS1-DX/FS1M-DX (2GO, 3F6) model forks differ in that they have no separate bushes; the inner tube bears directly on the fork lower leg. Also, a damper rod assembly complete with piston ring and rebound spring is fitted inside the inner

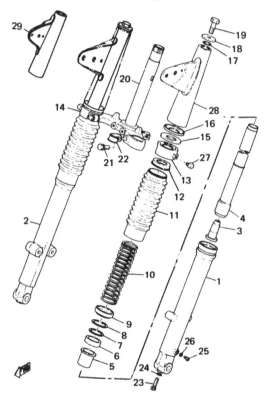

Fig. 7.11 Front forks – FS1E-DX (596) and FS1E-DXA [FS1 (2RV, 3AU) similar]

1 Left-hand lower leg	17 O-ring – 2 off
2 Right-hand lower leg	18 Cap washer – 2 off
3 Restrictor – 2 off*	19 Top bolt – 2 off
4 Inner tube – 2 off	20 Bottom yoke complete
5 Bush – 2 off	21 Pinch bolt – 2 off
6 Oil seal – 2 off	22 Wire holder
7 Washer – 2 off	23 Bolt – 2 off*
8 Circlip – 2 off	24 Sealing washer – 2 off*
9 Spring guide – 2 off	25 Drain plug – 2 off*
10 Fork spring – 2 off	26 Drain plug gasket – 2 off*
11 Gaiter – 2 off	27 Plug – 2 off
12 Upper spring seat – 2 off	28 Left-hand headlamp bracket
13 Left-hand outer cover	29 Right-hand headlamp
14 Right-hand outer cover	bracket
15 Packing – 2 off	
16 Under cover guide – 2 off	* FS1E-DX (596) and FS1E-DXA only

tube and is retained by an Allen screw which passes upwards through the bottom of the fork lower leg. The oil seal is mounted in the lower leg and is retained by a circlip but need not be disturbed during the course of dismantling unless it is to be renewed.

6   To dismantle the forks, unscrew the Allen screw and pull the stanchion (inner tube) out of the lower leg. It may be necessary to insert a length of wooden dowel inside the stanchion to bear down on the head of the damper rod, thus locking it while the screw is removed. To remove the oil seal, displace the circlip and lever the seal out of the lower leg, taking care not to scratch or distort the seal housing. Pouring boiling water over the lower leg will help to release a particularly tight

seal, but take great care to prevent personal injury. Invert the stanchion to tip out the damper rod assembly.

7   Reassembly is a straightforward reversal of the above. Tap the seal into its housing using a socket spanner or similar tubular drift which bears only on the seal outer edge, then refit the circlip. Place the damper rod assembly inside the stanchion and use the wooden dowel to restrain it while the Allen screw is tightened.

Steering head

8   Note that on all later models the steering head bottom bearing dust seal (item 1, Fig. 4.1) is no longer fitted.

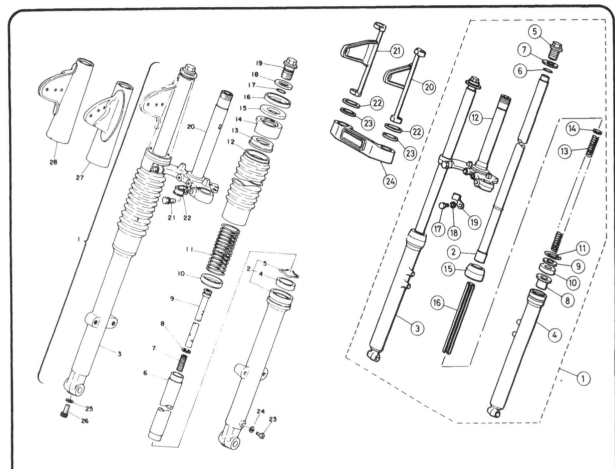

Fig. 7.12 Front forks – FS1-DX/FS1M-DX (2GO, 3F6)

1	Front fork assembly	15	Gasket – 2 off
2	Left-hand lower leg	16	Cover under guide – 2 off
3	Right-hand lower leg	17	Gasket – 2 off
4	Oil seal – 2 off	18	Washer – 2 off
5	Clip – 2 off	19	Bolt – 2 off
6	Inner tube – 2 off	20	Under bracket assembly
7	Rebound spring – 2 off	21	Bolt – 2 off
8	Front fork piston ring – 2 off	22	Cable holder
9	Damper tube – 2 off	23	Drain plug – 2 off
10	Spring guide – 2 off	24	Drain plug gasket – 2 off
11	Fork spring – 2 off	25	Gasket – 2 off
12	Gaiter – 2 off	26	Bolt – 2 off
13	Spring upper seat – 2 off	27	Upper left fork cover
14	Outer cover – 2 off	28	Upper right fork cover

Fig. 7.13 Front forks – FS1SE

1	Front fork assembly	15	Dust seal – 2 off
2	Inner tube – 2 off	16	Spacer – 2 off
3	Right-hand lower leg	17	Bolt – 2 off
4	Left-hand lower leg	18	Spring washer – 2 off
5	Top bolt – 2 off	19	Cable guide
6	O-ring – 2 off	20	Left-hand headlamp bracket
7	Cap washer – 2 off	21	Right-hand headlamp bracket
8	Bush – 2 off	22	Headlamp bracket guide – 2 off
9	Washer – 2 off	23	Packing – 2 off
10	Oil seal – 2 off	24	Bottom yoke cover
11	Circlip – 2 off		
12	Bottom yoke		
13	Fork spring – 2 off		
14	Washer – 2 off		

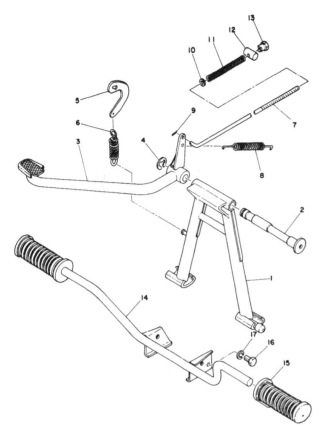

Fig. 7.14 Centre stand, footrests and brake pedal

1 Centre stand	10 Washer
2 Centre stand shaft	11 Compression spring
3 Brake pedal	12 Trunnion
4 Circlip	13 Nut
5 Centre stand link arm	14 Footrest
6 Tension spring	15 Footrest rubber – 2 off
7 Brake rod	16 Bolt
8 Tension spring	17 Spring washer
9 Split pin	

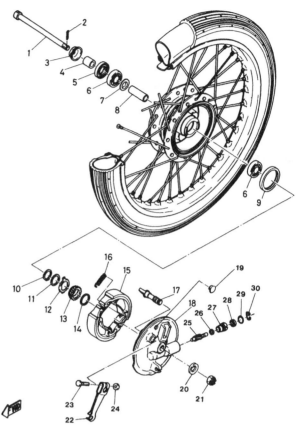

**Fig. 7.15 Front wheel and brake components –
FS1/FS1M (2GO, 2RV, 3AU) and FS1SE**

1 Wheel spindle	17 Brake operating cam
2 Split pin	18 Brake backplate
3 Dust seal	19 Inspection aperture plug
4 Spacer	20 Washer
5 Oil seal	21 Nut
6 Wheel bearing – 2 off	22 Brake operating arm
7 Spacer flange	23 Pinch bolt
8 Spacer	24 Nut
9 Oil seal	25 Speedometer driven
10 Circlip	pinion
11 Washer	26 Washer
12 Speedometer drive plate	27 Bush
13 Speedometer drive pinion	28 Oil seal
14 Washer	29 O-ring
15 Brake shoe – 2 off	30 Clip
16 Return spring – 2 off	

11 Front wheel – general – drum brake models

1   Note that on all later drum (front) brake models, the brake
backplate is now on the left-hand side of the machine. Refer to
the accompanying illustration for details of differences in
component layout.

2   Servicing and examination procedures are unchanged and
are given in Chapter 5 and Section 18 of this Chapter.

12 Front disc brake – pad renewal

FS1E-DX (596), FS1E-DXA, FS1-DX/FS1M-DX (2GO)

1   Place the machine on its centre stand and support it with a
wooden box underneath the crankcase so that the front wheel
is clear of the ground. Remove the front wheel. Remove the
single screw and plate which retain the fixed (outer) pad, then
tap the pad inwards to release it. The moving (inner) pad can be
displaced easily unless it is stuck by corrosion or road dirt. In
such cases, apply the front brake lever several times, using
hydraulic pressure to eject the pad, but note that under normal

circumstances this method should **not** be used or there is a risk
of ejecting the caliper piston.

2   Check that the caliper body slides easily from side to side;
if not, the caliper must be dismantled as described in Section 14
of this Chapter.

3   Reassembly is the reverse of the above. The pads are
different and cannot be interchanged; check that each is fitted
correctly in its housing. A thin smear of silicone- or PBC-based
brake caliper grease around the periphery of the moving pad
metal backing will prevent the onset of corrosion and preserve
braking efficiency. If new pads are to be fitted, or if the brake
lever was applied, the piston must be pressed back to provide

12.1a Remove single screw and support plate ...

12.1b ... to release fixed (outer) pad – FS1E-DX (596), FS1E-DXA and FS1-DX/FS1M-DX (2GO) models

sufficient clearance for the pads and disc to be refitted. If the piston cannot be pressed back easily, it is probable that corrosion has formed; the caliper must be dismantled to remedy this in the interests of braking efficiency. When the pads and front wheel have been refitted apply the brake lever several times until the pads are again in contact with the disc and full pressure has been restored. Check the level of fluid in the master cylinder reservoir; if new pads have been fitted, the level must be restored to the upper level mark, by topping up with clean fluid or by soaking away any surplus with a rag. If the original pads have been refitted, it is sufficient to ensure that the level is above the lower level mark. Refit the diaphragm and the cap or cover, tightening it securely.

FS1-DX (3F6)

4   Remove the two bolts retaining the caliper to the fork lower leg and lift the caliper off the disc, taking care not to distort the brake hose. Each pad is retained in an anti-squeal shim by a thin spring clip. Prise out the clips and displace the pads, noting which way round each is fitted. The moving pad may be dislodged by careful leverage, but in severe cases the brake lever can be applied to use hydraulic pressure to displace the pad. Note that in normal circumstances this method should **not** be used or there is a risk of ejecting the caliper piston.

5   Check that the caliper body rotates smoothly about its pivot; if not, the caliper must be dismantled as described in Section 14 for cleaning and examination.

6   On reassembly, refit each anti-rattle shim to its correct pad followed by the spring clips. Ensure that each pad is refitted in its correct location and is securely retained. If new pads have been fitted, or if the brake lever was applied, the piston must be pressed back to provide the necessary clearance. If the piston cannot be pressed back easily it is probable that corrosion has formed; the caliper must be dismantled immediately for cleaning and examination. Refit the caliper to the machine, tightening its two mounting bolts to a torque setting of 2.0 kgf m (14.5 lbf ft).

7   Apply the front brake lever several times until the pads are in firm contact with the disc and full braking pressure has been restored, then check the fluid level in the master cylinder reservoir. If new pads have been fitted, the level must be restored to the upper level mark, either by topping up with new fluid or by soaking away the surplus with a clean rag, but if the original pads have been refitted, it is sufficient to ensure that the level is above the lower level mark. Refit the diaphragm and cover, tightening securely the retaining screws.

All models

8·   If, on examination, the brake pads are found to be chipped,

damaged, fouled with oil or grease, or excessively worn, they must be renewed. On the early type of caliper the friction material wear limit is indicated by a step in the material next to the metal backing or by a red-painted band; if the marks are not obvious, measure the total thickness of the pad, friction material and metal backing. If either pad is worn to the limit mark, or to the thickness limit given in the Specifications Section of this Chapter, both must be renewed as a matter of course. On the later type of caliper the wear limit is shown by projecting tongues extending from each anti-rattle shim; if these come into contact with the disc (immediately obvious due to the noise produced) the pads must be renewed. Note that the shims should also be renewed if they are found to be seriously worn or distorted.

9   If the friction material is found to be in good condition and within wear limits, use a clean wire brush to scrub off all traces of road dirt and corrosion, both from the friction material and the metal backing. Remove any embedded particles from the rubbing surface and remove any traces of glazing by rubbing carefully with emery paper to break the glazed surface.

13 Front brake disc – examination and renovation

1   The disc can be examined while in place on the machine. If it is heavily scored or worn at any point to less than the limit specified, it must be renewed. If signs of warpage can be seen by the naked eye on spinning the front wheel, a dial gauge must be obtained and mounted on the fork lower leg to measure accurately the amount of distortion. If this proves excessive, the disc must be renewed. The only alternative is to find an engineering company who will skim the disc, but this must not reduce it to less than the minimum thickness specified, or braking efficiency will be greatly reduced, quite apart from the risk of sudden brake failure.

2   If the disc is to be renewed, remove the front wheel, knock back the raised tabs of the lock washers and remove the four disc mounting bolts. On reassembly, tighten the bolts evenly to the specified torque setting and secure each one by bending up an unused tab against one of the flats of its head.

14 Front brake caliper – examination and renovation

1   For both types of caliper, commence operations by removing the pads as described in Section 12 of this Chapter. Note that while the usual method of caliper piston removal is described below, it is equally effective to apply the brake lever

14.3 FS1E-DX (596), FS1E-DXA, FS1-DX/FS1M-DX (2GO) caliper – remove two mounting bolts to release

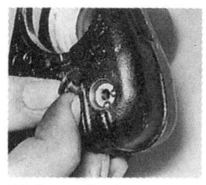

14.5a Remove rubber plugs ...

14.5b ... withdraw circlips ...

14.5c ... and extract caliper axle pins as shown

14.5d Caliper support bracket can then be removed from caliper body

14.5e Note which way round anti-rattle spring is fitted – do not omit

to displace the piston using hydraulic pressure before the brake hoses are disconnected. If this method is employed, wrap the caliper in a plastic bag to catch the piston and to restrict the resulting shower of brake fluid.

2   Note that only the pads and the caliper piston and seals can be purchased as separate replacement parts; if severe wear or damage of any sort is encountered, the complete caliper assembly must be renewed.

FS1E-DX (596), FS1E-DXA, FS1-DX/FS1M-DX (2GO)

3   Before the caliper assembly can be removed from the right-hand fork leg, it is necessary to drain off the hydraulic fluid. Disconnect the brake pipe at the union connection it makes with the caliper unit and allow the fluid to drain into a clean container. Keep the front brake lever applied throughout this operation, to prevent the fluid from leaking out of the reservoir. A thick rubber band cut from a section of inner tube will suffice, it if is wrapped tightly around the lever and the handlebars. Cover the end of the pipe with a polythene bag, after the fluid has drained, to keep it clean. Remove the two caliper mounting bolts and lift away the caliper.

4   Brake fluid is an extremely efficient paint stripper. Take care to keep it away from any paintwork on the machine or from any clear plastic, such as that sometimes used for instrument glasses.

5   Remove the ring that retains the rubber boot, the boot itself, and the two blind plugs in the side of the unit, which can be prised out with a screwdriver. Remove the small circlips, then withdraw the two pins that retain the inner end of the caliper support bracket. They are threaded internally and can be withdrawn with a pair of pliers if a 5 mm screw is inserted into the threaded portion. Remove the support bracket and the anti-rattle spring. Check carefully for signs of wear on the pins and bracket.

6   To displace the piston, apply a blast of compressed air through the brake fluid inlet. Take care to catch the piston as it

emerges from its bore – if dropped or prised out of position with a screwdriver, it may be damaged irreparably and will have to be renewed. Remove the piston seal and dust seal from the caliper body.

7   The parts removed should be cleaned thoroughly in new brake fluid. Petrol, oil or paraffin will cause the various seals to swell and degrade. When the various parts have been cleaned, they should be stored in polythene bags until reassembly, so that they are kept dust free.

8   The piston should be carefully examined for scratches, score marks, and other such imperfections. If any are found the

14.6 Use hydraulic pressure or compressed air to displace caliper piston

piston should be replaced, otherwise air or hydraulic fluid leakage will occur which will impair braking efficiency. With regard to the various seals, it is advisable to renew them irrespective of their appearance. It is a small price to pay against the risk of a sudden complete front brake failure. It is standard Yamaha practice to renew the seals every two years, even if no braking problems have occurred.

9 Reassembly under clinically clean conditions, by reversing the dismantling procedure. Apply caliper grease to the surfaces of the pins and the sliding surfaces of the bracket and body. Reconnect the hydraulic fluid pipe and make sure the union has been tightened fully. Before the brake can be used, the whole system must be bled of air, by following the procedure described in Section 17 of this Chapter.

FS1-DX (3F6)

10 The procedure for removing and checking the caliper piston and seals is the same as that described in paragraphs 6, 7 and 8 above.

11 The major difference between the two types of caliper is in the method employed to allow the caliper body to swing so that equal pressure is applied to both faces of the disc when braking. While the body of the earlier type slides across on two pins bearing on the fixed support bracket, this later type pivots about an axis formed by an extension of its mounting bracket. This is where most problems are likely to be found, due to the corrosion formed by rainwater, road dirt, salt etc.

12 Remove the split pin, retaining nut and the plain washer then slide the caliper body off its pivot. Carefully remove all traces of dirt and corrosion, then refit the body and feel for signs of free play. If excessive wear is found, the caliper assembly must be renewed as a complete unit.

13 On reassembly, apply a thin smear of silicone- or PBC-based caliper grease to the surface of the pivot and to the bore of the caliper body, refit the body followed by the plain washer, tighten securely the retaining nut, then check that the body is free to pivot and refit the split pin, spreading its ends securely.

14 Reconnect the hydraulic pipe and bleed the system as described in Section 17.

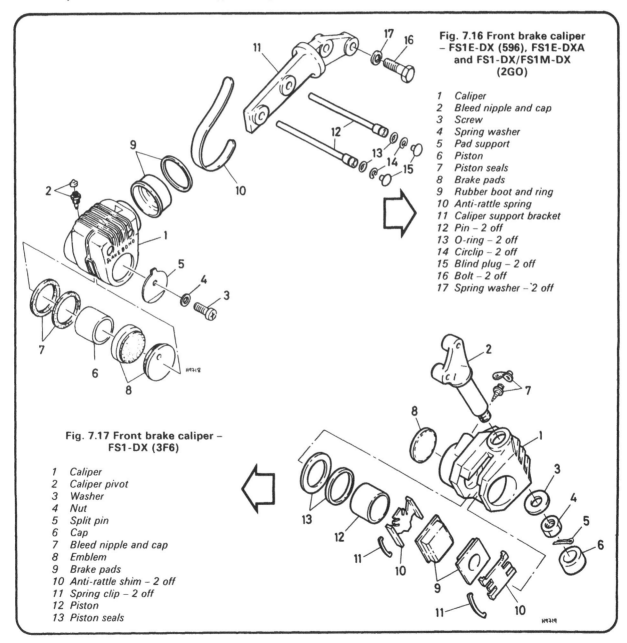

Fig. 7.16 Front brake caliper – FS1E-DX (596), FS1E-DXA and FS1-DX/FS1M-DX (2GO)

1   Caliper
2   Bleed nipple and cap
3   Screw
4   Spring washer
5   Pad support
6   Piston
7   Piston seals
8   Brake pads
9   Rubber boot and ring
10 Anti-rattle spring
11 Caliper support bracket
12 Pin – 2 off
13 O-ring – 2 off
14 Circlip – 2 off
15 Blind plug – 2 off
16 Bolt – 2 off
17 Spring washer – 2 off

Fig. 7.17 Front brake caliper – FS1-DX (3F6)

1   Caliper
2   Caliper pivot
3   Washer
4   Nut
5   Split pin
6   Cap
7   Bleed nipple and cap
8   Emblem
9   Brake pads
10 Anti-rattle shim – 2 off
11 Spring clip – 2 off
12 Piston
13 Piston seals

15 Master cylinder – examination and renovation

1   The master cylinder and hydraulic fluid reservoir take the form of a combined unit mounted on the right-hand side of the handlebars, to which the front brake lever is attached. Note that while there are two types of master cylinder that are markedly different in appearance to the third, the actual differences between the working parts of the three types are minor. Study carefully the appropriate illustration before commencing work.
2   Before the master cylinder unit can be removed and dismantled, the system must be drained. Place a clean container below the brake caliper unit and attach a plastic tube from the bleed screw of the caliper unit to the container. Open the bleed screw one complete turn and drain the system by operating the brake lever repeatedly until the master cylinder reservoir is empty. Close the bleed screw and remove the tube.
3   Before dismantling the master cylinder, it is essential that a clean working area is available on which the various component parts can be laid out. Use a sheet of white paper, so that none of the smaller parts can be overlooked.
4   Disconnect the stop lamp switch and front brake lever, taking care not to misplace the brake lever return spring. The stop lamp switch is attached to the bolt that acts as the brake lever pivot or is set in the master cylinder body. Remove the split pin and castellated nut and take off the switch or disconnect it and remove it with the master cylinder. Remove the brake hose by unscrewing the banjo union bolt, then remove the two clamp bolts and withdraw the assembly from the handlebars. Check that all fluid has drained from the reservoir by taking off the reservoir cap or cover and the diaphragm below.
5   Withdraw the rubber boot that protects the end of the master cylinder, noting that this may be retained by a circlip and remove the circlip that holds the piston assembly in position, using a pair of circlip pliers. The piston assembly and spring can now be withdrawn.

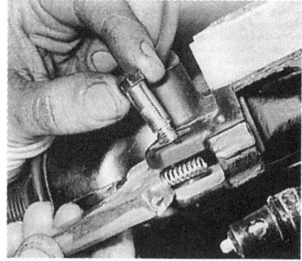

15.4a Do not omit brake lever return spring – note later type of stop lamp switch

6   To dismantle the piston assembly, where applicable, remove the 'E' clip at the far end and then the primary cup retainer. The primary cup can then be detached.
7   Examine the piston and the primary cup very carefully. If either is scratched or has the working surface impaired in any other way, it must be renewed without question. Reject the various seals, irrespective of their condition, and fit new ones in their place. It often helps to soften them a little before they are fitted by immersing them in a container of clean brake fluid.
8   When reassembling, follow the dismantling procedure in reverse, but take great care that none of the component parts

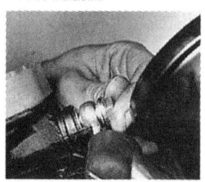

15.4b Remove banjo bolt to release brake hose – renew sealing washers on reassembly

15.5a Rubber boot is retained by a circlip on some models

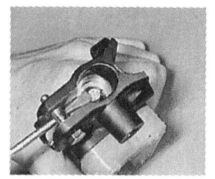

15.5b If necessary, lever piston inwards so that circlip can be removed ...

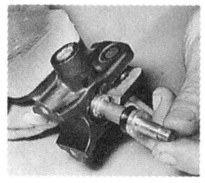

15.5c ... and the piston assembly withdrawn

15.5d Note which way round components are fitted before removing

Chapter 7 The 1975 on models

113

are scratched or damaged in any way. Use brake fluid as the lubricant whilst reassembling. When assembly is complete, reconnect the brake fluid pipe and tighten the banjo union bolt

to the recommended setting. Refill the system with new brake fluid of the recommended type and remove all traces of air by bleeding the system as described in Section 17 of this Chapter.

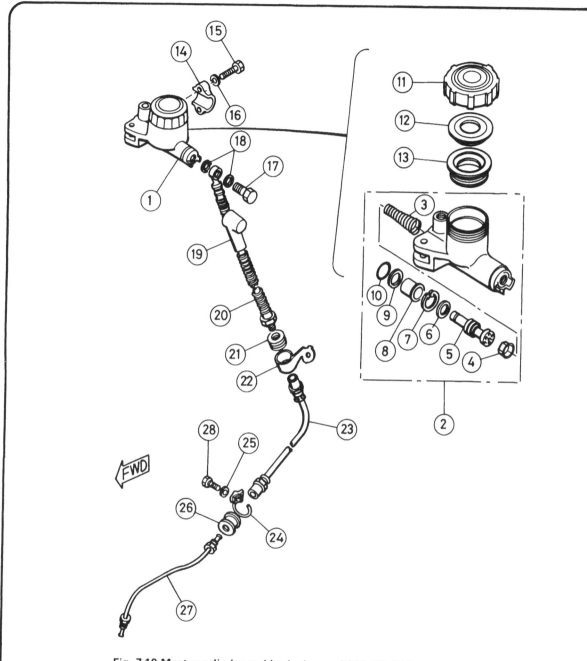

Fig. 7.18 Master cylinder and brake hose – FS1E-DX (596) and FS1E-DXA

1	Master cylinder assembly	11	Reservoir cap
2	Master cylinder kit	12	Reservoir diaphragm plate
3	Spring	13	Reservoir diaphragm
4	Primary cup	14	Handlebar clamp
5	Piston	15	Bolt – 2 off
6	Washer	16	Spring washer – 2 off
7	Circlip	17	Banjo union bolt
8	Rubber boot	18	Sealing washer – 2 off
9	Washer	19	Rubber boot
10	Circlip		

20	Upper brake hose
21	Grommet
22	Clamp
23	Lower brake hose
24	Clamp
25	Spring washer
26	Grommet
27	Metal brake pipe
28	Bolt

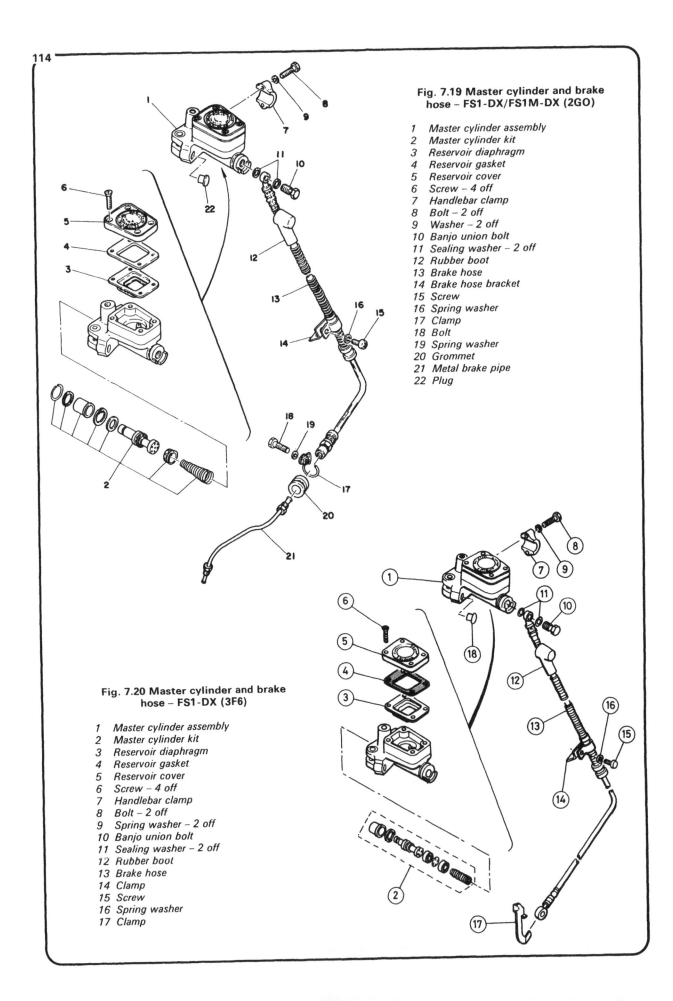

Fig. 7.19 Master cylinder and brake hose – FS1-DX/FS1M-DX (2GO)

1 Master cylinder assembly
2 Master cylinder kit
3 Reservoir diaphragm
4 Reservoir gasket
5 Reservoir cover
6 Screw – 4 off
7 Handlebar clamp
8 Bolt – 2 off
9 Washer – 2 off
10 Banjo union bolt
11 Sealing washer – 2 off
12 Rubber boot
13 Brake hose
14 Brake hose bracket
15 Screw
16 Spring washer
17 Clamp
18 Bolt
19 Spring washer
20 Grommet
21 Metal brake pipe
22 Plug

Fig. 7.20 Master cylinder and brake hose – FS1-DX (3F6)

1 Master cylinder assembly
2 Master cylinder kit
3 Reservoir diaphragm
4 Reservoir gasket
5 Reservoir cover
6 Screw – 4 off
7 Handlebar clamp
8 Bolt – 2 off
9 Spring washer – 2 off
10 Banjo union bolt
11 Sealing washer – 2 off
12 Rubber boot
13 Brake hose
14 Clamp
15 Screw
16 Spring washer
17 Clamp

16 Hydraulic brake hose and pipe – examination

1 An external brake hose and pipe is used to transmit the hydraulic pressure to the caliper unit when the front brake is applied. The brake hose is of the flexible type, fitted with an armoured surround. It is capable of withstanding pressures up to 350 kg/cm^2. The brake pipe attached to it on early models is made from double steel tubing, zinc plated to give better corrosion resistance.

2 When the brake assembly is being overhauled, check the condition of both the hose and the pipe for signs of leakage or scuffing. The union connections at either end must also be in good condition, with no stripped threads or damaged sealing washers. Renew any hose or pipe immediately if found to be worn or damaged, referring to the torque settings given in the Specifications Section of this Chapter. Also renew any sealing washers whenever they are disturbed.

17 Hydraulic brake system – bleeding

1 If the brake action becomes spongy, or if any part of the hydraulic system is dismantled (such as when a hose is replaced) it is necessary to bleed the system in order to remove all traces of air. The procedure for bleeding the hydraulic system is best carried out by two people.

2 Check the fluid level in the reservoir and top up with new fluid of the specified type if required. Keep the reservoir at least half full during the bleeding procedure; if the level is allowed to fall too far, air will enter the system requiring that the procedure be started again from scratch. Replace the reservoir cap or cover to prevent the ingress of dust or the ejection of a spout of fluid.

3 Remove the dust cap from the caliper bleed nipple and clean the area with a rag. Place a clean glass jar below the caliper and connect a pipe from the bleed nipple to the jar. A clear plastic tube should be used so that air bubbles can be more easily seen. Place some clean hydraulic fluid in the glass jar so that the pipe is immersed below the fluid surface throughout the operation.

4 If parts of the system have to be renewed, and thus the system must be filled, open the bleed nipple about one turn and pump the brake lever until fluid starts to issue from the clear tube. Tighten the bleed nipple and then continue the normal bleeding operation as described in the following paragraphs. Keep a close check on the reservoir level whilst the system is being filled.

5 Operate the brake lever as far as it will go and hold it in this position against the fluid pressure. If spongy brake operation has occurred, it may be necessary to pump the brake lever rapidly a number of times until pressure is achieved. With pressure applied, loosen the bleed nipple about half a turn. Tighten the nipple as soon as the lever has reached its full travel and then release the lever. Repeat this operation until no more air bubbles are expelled with the fluid into the glass jar. When this condition is reached the air bleeding operation should be complete, resulting in a firm feel to the brake operation. If sponginess is still evident, continue the bleeding operation; it may be that an air bubble trapped at the top of the system has yet to work through the caliper.

6 When all traces of air have been removed from the system, top up the reservoir and refit the diaphragm and cap or cover, as appropriate. Check the entire system for leaks, and check also that the brake system in general is functioning efficiently before using the machine on the road.

7 Brake fluid drained from the system will almost certainly be contaminated, either by foreign matter or more commonly by the absorption of water from the air. All hydraulic fluids are to some degree hygroscopic, that is, they are capable of drawing water from the atmosphere, and thereby degrading their specifications. In view of this, and the relative cheapness of the fluid, old fluid should always be discarded.

8 Great care should be taken not to spill hydraulic fluid on any painted cycle parts; it is a very effective paint stripper. Also, the plastic glasses in the instrument heads, and most other plastic parts, will be damaged by contact with this fluid.

18 Brake shoe wear – check

All models

1 The simplest method of assessing the amount of wear that has taken place on the brake shoe friction material is to check the angle formed between the brake operating arm and the cable (front) or rod (rear) when the brake is correctly adjusted and fully applied.

2 If the brake components are in good condition the angle will be less than 90°. If the angle is found to be greater than 90° at any time the wheel must be removed from the machine as soon as possible and the brake components should be dismantled, cleaned and checked for wear. Any worn components should be renewed. Note: *when cleaning the brake components and wheel hub, be careful to take suitable precautions to prevent the risk of inhaling any of the asbestos dust – see Safety first!, page 9.*

3 If the brake shoes are still fit for further use, the operating arm can be removed from the cam splines and rotated through one or two splines until the angle is correct; be careful to ensure that the arm is fully secure on the cam splines on refitting, and that the pinch bolt is securely fastened. Note, however, that if the arm is moved an excessive amount it will no longer be possible to use this method to determine brake shoe wear; instead it will be necessary to remove the wheel at each service interval to permit a full check of the brake components. There are, however, alternative methods of checking brake shoe wear provided by the manufacturer, as described below.

FS1SE

4 This model is fitted with small pointers which are positioned on each brake cam, inboard of the operating arm, to register on an arc cast into each brake backplate. With the brake correctly adjusted and fully applied, if the pointer moves beyond the wear limit line at the end of the arc the brake shoes are worn out and must be renewed.

FS1/FS1M (2GO), FS1-DX/FS1M-DX (2GO, 3F6) and FS1 (2RV, 3AU)

5 These models are fitted with inspection apertures in their brake backplates which permit the friction material of one brake shoe to be checked by direct measurement without removing the wheel from the machine. These apertures are sealed by black rubber plugs (see photo 1i).

6 The friction material can be easily measured by inserting a rod of 2 mm (0.08 in) diameter into the aperture and comparing the thickness of the rod with that of the lining. If the friction material is worn to the diameter of the rod or less, at any point on the shoe, or even if it is approaching this thickness, the wheel must be removed from the machine so that the brake shoes can be fully and accurately checked. Remember that brake shoes do not wear evenly and must be renewed as a pair if *either* is worn *at any point* to a thickness of 2.0 mm (0.08 in) or less.

7 Always ensure that the plugs are securely refitted after each check to prevent the entry of dirt or water.

19 Rear wheel – general

1 All restricted models are fitted with a modified rear wheel in which the sprocket is bolted directly to the wheel hub; the cush drive shock absorber assembly described in Chapter 5 is no longer fitted. Refer to the accompanying illustration for details.

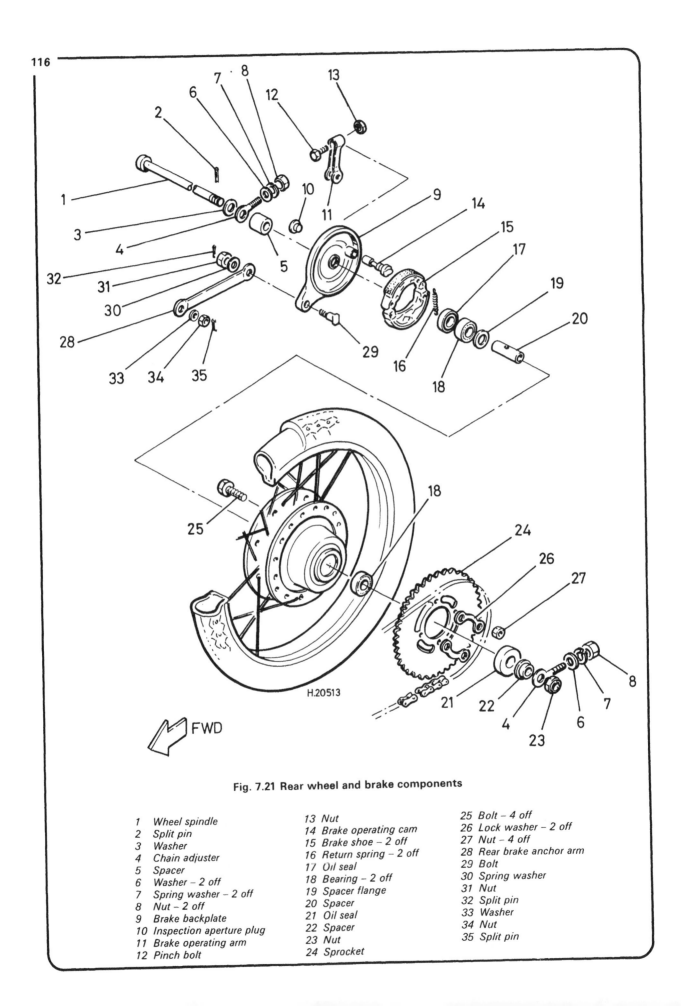

Fig. 7.21 Rear wheel and brake components

1	Wheel spindle	13	Nut	25	Bolt – 4 off
2	Split pin	14	Brake operating cam	26	Lock washer – 2 off
3	Washer	15	Brake shoe – 2 off	27	Nut – 4 off
4	Chain adjuster	16	Return spring – 2 off	28	Rear brake anchor arm
5	Spacer	17	Oil seal	29	Bolt
6	Washer – 2 off	18	Bearing – 2 off	30	Spring washer
7	Spring washer – 2 off	19	Spacer flange	31	Nut
8	Nut – 2 off	20	Spacer	32	Split pin
9	Brake backplate	21	Oil seal	33	Washer
10	Inspection aperture plug	22	Spacer	34	Nut
11	Brake operating arm	23	Nut	35	Split pin
12	Pinch bolt	24	Sprocket		

2 The wheel removal procedure remains much as described in Chapter 5 although due to the modified design it is now necessary to first disconnect the final drive chain. The spindle can then be removed and the wheel withdrawn, complete with the sprocket and brake assembly. Remember to check the chain tension and adjust it if necessary on refitting.

3 On adjusting the chain, it is now necessary to slacken only the wheel spindle nut to permit the adjusters to be moved.

4 Note that the rear sprocket mountings are similar to those previously described, with the bolts or nuts being fitted from inside the brake drum.

20 Higher charging rate modifications

1 Some machines have suffered their batteries being continually discharged due to excessive slow-speed riding or to constant use of the flashing indicators and stop lamp. This applies mainly to the standard models, ie the FS1E (394), FS1E-A, FS1/FS1M (2GO), but may be encountered also on DX models.

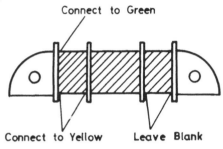

Connect to Green

Connect to Yellow   Leave Blank

Fig. 7.22 Connections for higher output generator coil – FS1/FS1M (2GO) only

2 For the FS1E (394) and FS1E-A models the answer is in an alternative connection for the generator source coil wires, and is shown adjacent to the relevant wiring diagram.

3 For FS1/FS1M (2GO) models a modified higher output coil is available under part number 594-81313-20. This is connected to the existing wires as shown in the accompanying illustration. Note that this was carried out under warranty in severe cases, and so may already be fitted to the machine being worked on.

21 Voltage regulator – FS1SE

1 This is a small sealed unit mounted inside the frame on the rear of the battery carrier, just beneath the flashing indicator relay. Its purpose is to soak up excess current generated when the lighting switch is in the 'Off' position and to prevent bulb blowing due to voltage surges.

2 Since there is no information to assist the owner to test the unit, it can be checked only by the substitution of a new component. Check first all switches for correct operation, that all bulbs are of the correct type and rating, and that the wiring and connections, including earth points and bulb contacts, are in good condition.

22 Flashing indicators – FS1 (2RV, 3AU)

Note that the square-bodied flashing indicator lamp assemblies use two different screws to secure each lens; ensure that the screws are refitted correctly on reassembly. The flashing indicator relay is a black plastic square-bodied unit mounted behind and to the rear of the battery.

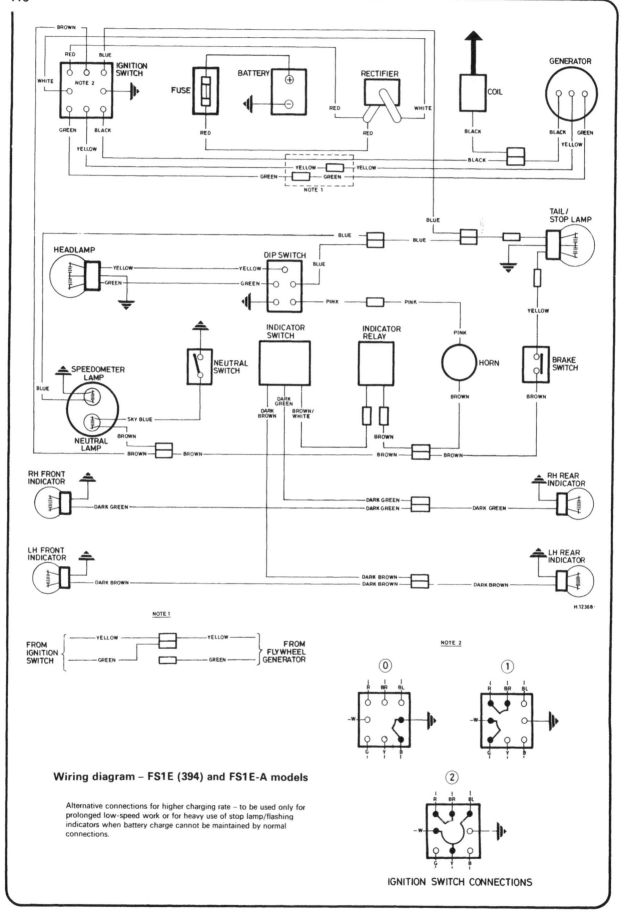

Wiring diagram – FS1E (394) and FS1E-A models

Alternative connections for higher charging rate – to be used only for
prolonged low-speed work or for heavy use of stop lamp/flashing
indicators when battery charge cannot be maintained by normal
connections.

IGNITION SWITCH CONNECTIONS

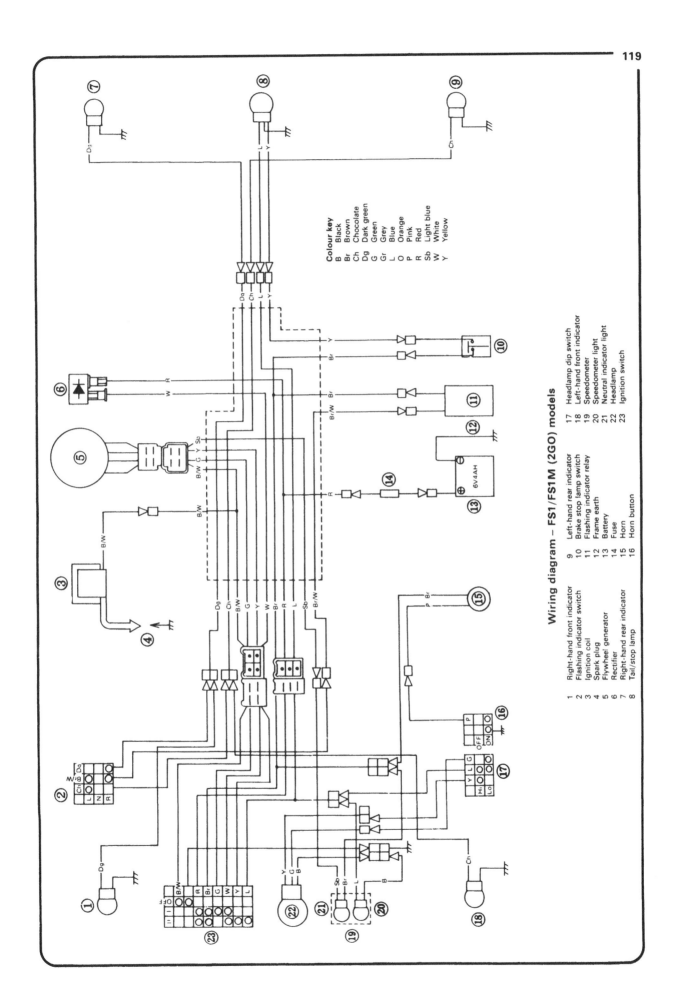

Colour key

B	Black
Br	Brown
Ch	Chocolate
Dg	Dark green
G	Green
Gr	Grey
L	Blue
O	Orange
P	Pink
R	Red
Sb	Light blue
W	White
Y	Yellow

Wiring diagram – FS1/FS1M (2GO) models

1	Right-hand front indicator	9	Left-hand rear indicator	17 Headlamp dip switch
2	Flashing indicator switch	10	Brake stop lamp switch	18 Left-hand front indicator
3	Ignition coil	11	Flashing indicator relay	19 Speedometer
4	Spark plug	12	Frame earth	20 Speedometer light
5	Flywheel generator	13	Battery	21 Neutral indicator light
6	Rectifier	14	Fuse	22 Headlamp
7	Right-hand rear indicator	15	Horn	23 Ignition switch
8	Tail/stop lamp	16	Horn button	

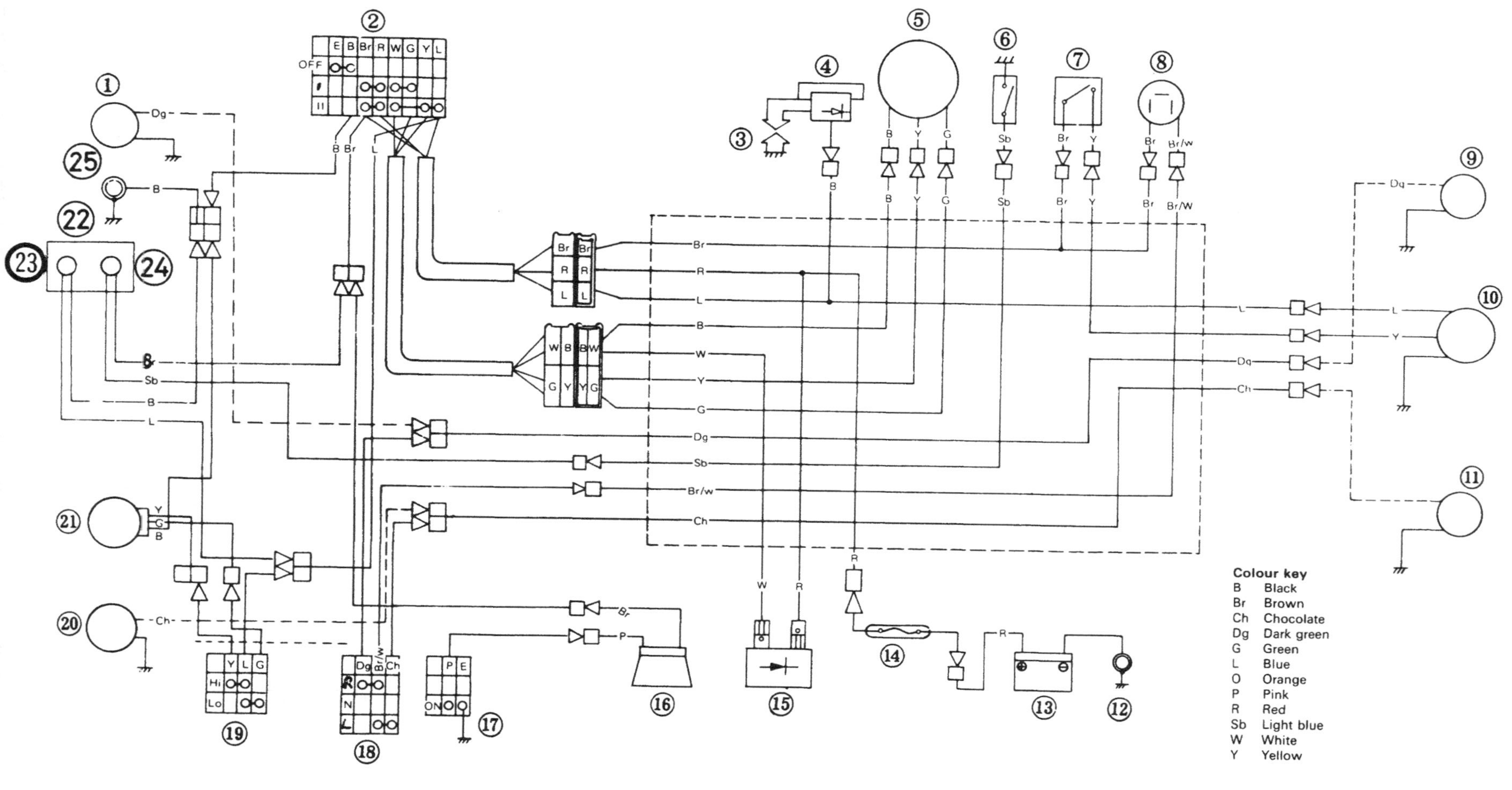

Colour key

B	Black
Br	Brown
Ch	Chocolate
Dg	Dark green
G	Green
L	Blue
O	Orange
P	Pink
R	Red
Sb	Light blue
W	White
Y	Yellow

Wiring diagram – FS1-DX/FS1M-DX (2GO) models

1 Right-hand front indicator
2 Ignition switch
3 Spark plug
4 Ignition coil
5 Flywheel generator
6 Neutral indicator switch
7 Brake stop lamp switch
8 Flashing indicator relay
9 Right-hand rear indicator
10 Tail/stop lamp
11 Left-hand rear indicator
12 Earth point
13 Battery
14 Fuse
15 Rectifier
16 Horn
17 Horn button
18 Flashing indicator switch
19 Headlamp dip switch
20 Left-hand front indicator
21 Headlamp
22 Speedometer
23 Speedometer light
24 Neutral indicator light
25 Earth point

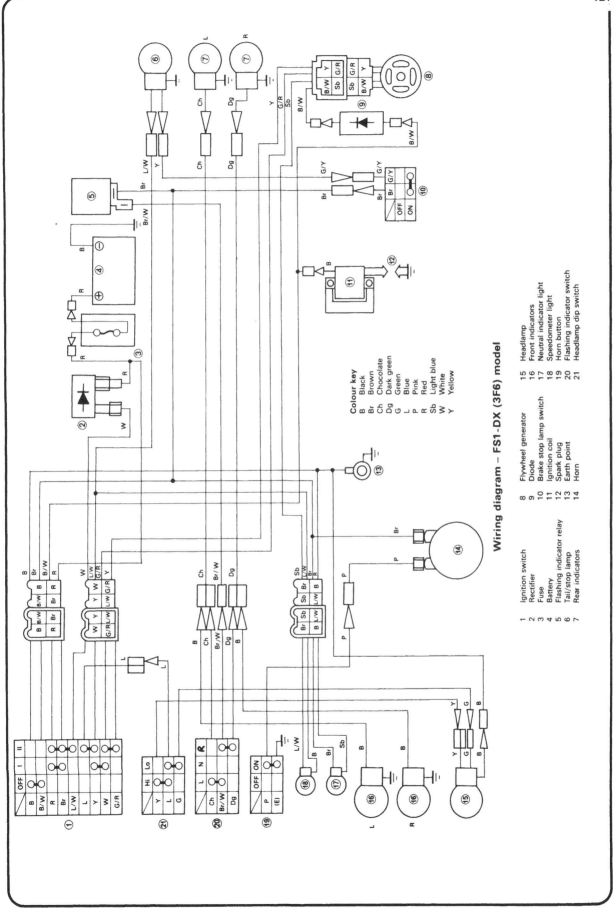

Wiring diagram – FS1-DX (3F6) model

Colour key

B	Black
Br	Brown
Ch	Chocolate
Dg	Dark green
G	Green
L	Blue
P	Pink
R	Red
Sb	Light blue
W	White
Y	Yellow

1	Ignition switch	8	Flywheel generator
2	Rectifier	9	Diode
3	Fuse	10	Brake stop lamp switch
4	Battery	11	Ignition coil
5	Flashing indicator relay	12	Spark plug
6	Tail/stop lamp	13	Earth point
7	Rear indicators	14	Horn
		15	Headlamp
		16	Front indicators
		17	Neutral indicator light
		18	Speedometer light
		19	Horn button
		20	Flashing indicator switch
		21	Headlamp dip switch

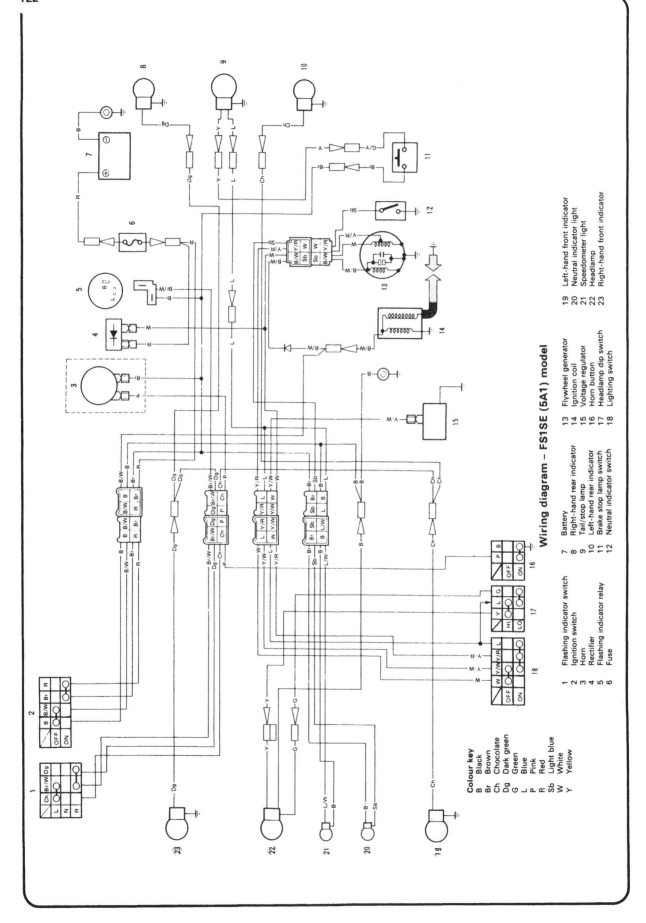

Wiring diagram – FS1SE (5A1) model

Colour key

B	Black
Br	Brown
Ch	Chocolate
Dg	Dark green
G	Green
L	Blue
P	Pink
R	Red
Sb	Light blue
W	White
Y	Yellow

1 Flashing indicator switch
2 Ignition switch
3 Horn
4 Rectifier
5 Flashing indicator relay
6 Fuse

7 Battery
8 Right-hand rear indicator
9 Tail/stop lamp
10 Left-hand rear indicator
11 Brake stop lamp switch
12 Neutral indicator switch

13 Flywheel generator
14 Ignition coil
15 Voltage regulator
16 Horn button
17 Headlamp dip switch
18 Lighting switch

19 Left-hand front indicator
20 Neutral indicator light
21 Speedometer light
22 Headlamp
23 Right-hand front indicator

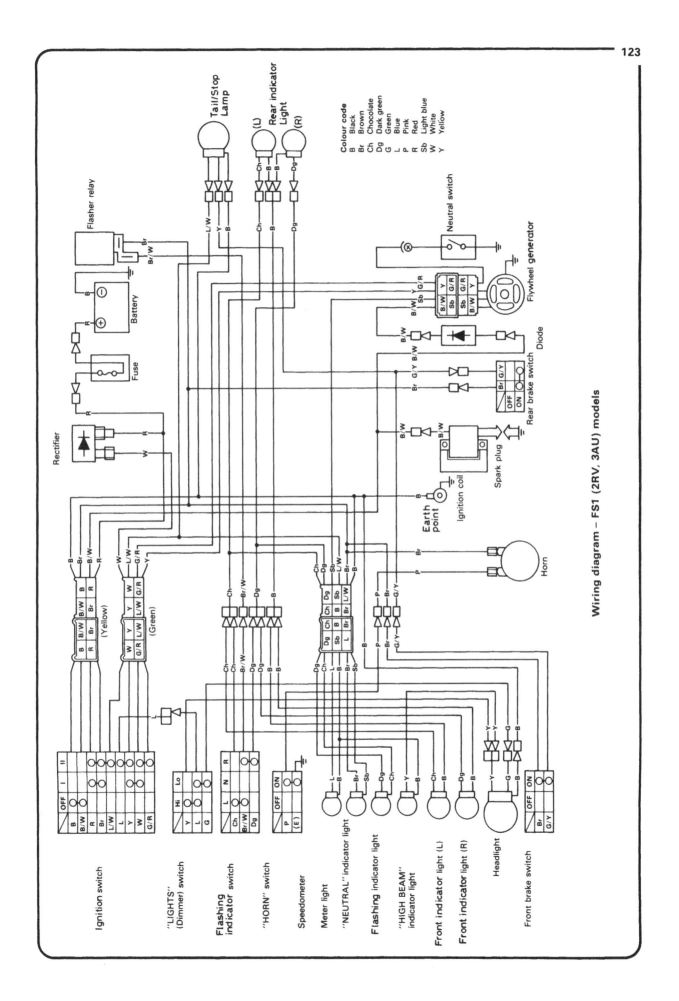

Wiring diagram – FS1 (2RV, 3AU) models

Conversion factors

Length (distance)

Inches (in)	X	25.4	= Millimetres (mm)	X 0.0394	= Inches (in)
Feet (ft)	X	0.305	= Metres (m)	X 3.281	= Feet (ft)
Miles	X	1.609	= Kilometres (km)	X 0.621	= Miles

Volume (capacity)

Cubic inches (cu in; in³)	X	16.387	= Cubic centimetres (cc; cm³)	X 0.061	= Cubic inches (cu in; in³)
Imperial pints (Imp pt)	X	0.568	= Litres (l)	X 1.76	= Imperial pints (Imp pt)
Imperial quarts (Imp qt)	X	1.137	= Litres (l)	X 0.88	= Imperial quarts (Imp qt)
Imperial quarts (Imp qt)	X	1.201	= US quarts (US qt)	X 0.833	= Imperial quarts (Imp qt)
US quarts (US qt)	X	0.946	= Litres (l)	X 1.057	= US quarts (US qt)
Imperial gallons (Imp gal)	X	4.546	= Litres (l)	X 0.22	= Imperial gallons (Imp gal)
Imperial gallons (Imp gal)	X	1.201	= US gallons (US gal)	X 0.833	= Imperial gallons (Imp gal)
US gallons (US gal)	X	3.785	= Litres (l)	X 0.264	= US gallons (US gal)

Mass (weight)

Ounces (oz)	X	28.35	= Grams (g)	X 0.035	= Ounces (oz)
Pounds (lb)	X	0.454	= Kilograms (kg)	X 2.205	= Pounds (lb)

Force

Ounces-force (ozf; oz)	X	0.278	= Newtons (N)	X 3.6	= Ounces-force (ozf; oz)
Pounds-force (lbf; lb)	X	4.448	= Newtons (N)	X 0.225	= Pounds-force (lbf; lb)
Newtons (N)	X	0.1	= Kilograms-force (kgf; kg)	X 9.81	= Newtons (N)

Pressure

Pounds-force per square inch (psi; lbf/in²; lb/in²)	X	0.070	= Kilograms-force per square centimetre (kgf/cm²; kg/cm²)	X 14.223	= Pounds-force per square inch (psi; lbf/in²; lb/in²)
Pounds-force per square inch (psi; lbf/in²; lb/in²)	X	0.068	= Atmospheres (atm)	X 14.696	= Pounds-force per square inch (psi; lbf/in²; lb/in²)
Pounds-force per square inch (psi; lbf/in²; lb/in²)	X	0.069	= Bars	X 14.5	= Pounds-force per square inch (psi; lbf/in²; lb/in²)
Pounds-force per square inch (psi; lbf/in²; lb/in²)	X	6.895	= Kilopascals (kPa)	X 0.145	= Pounds-force per square inch (psi; lbf/in²; lb/in²)
Kilopascals (kPa)	X	0.01	= Kilograms-force per square centimetre (kgf/cm²; kg/cm²)	X 98.1	= Kilopascals (kPa)
Millibar (mbar)	X	100	= Pascals (Pa)	X 0.01	= Millibar (mbar)
Millibar (mbar)	X	0.0145	= Pounds-force per square inch (psi; lbf/in²; lb/in²)	X 68.947	= Millibar (mbar)
Millibar (mbar)	X	0.75	= Millimetres of mercury (mmHg)	X 1.333	= Millibar (mbar)
Millibar (mbar)	X	0.401	= Inches of water (inH₂O)	X 2.491	= Millibar (mbar)
Millimetres of mercury (mmHg)	X	0.535	= Inches of water (inH₂O)	X 1.868	= Millimetres of mercury (mmHg)
Inches of water (inH₂O)	X	0.036	= Pounds-force per square inch (psi; lbf/in²; lb/in²)	X 27.68	= Inches of water (inH₂O)

Torque (moment of force)

Pounds-force inches (lbf in; lb in)	X	1.152	= Kilograms-force centimetre (kgf cm; kg cm)	X 0.868	= Pounds-force inches (lbf in; lb in)
Pounds-force inches (lbf in; lb in)	X	0.113	= Newton metres (Nm)	X 8.85	= Pounds-force inches (lbf in; lb in)
Pounds-force inches (lbf in; lb in)	X	0.083	= Pounds-force feet (lbf ft; lb ft)	X 12	= Pounds-force inches (lbf in; lb in)
Pounds-force feet (lbf ft; lb ft)	X	0.138	= Kilograms-force metres (kgf m; kg m)	X 7.233	= Pounds-force feet (lbf ft; lb ft)
Pounds-force feet (lbf ft; lb ft)	X	1.356	= Newton metres (Nm)	X 0.738	= Pounds-force feet (lbf ft; lb ft)
Newton metres (Nm)	X	0.102	= Kilograms-force metres (kgf m; kg m)	X 9.804	= Newton metres (Nm)

Power

Horsepower (hp)	X	745.7	= Watts (W)	X 0.0013	= Horsepower (hp)

Velocity (speed)

Miles per hour (miles/hr; mph)	X	1.609	= Kilometres per hour (km/hr; kph)	X 0.621	= Miles per hour (miles/hr; mph)

Fuel consumption*

Miles per gallon, Imperial (mpg)	X	0.354	= Kilometres per litre (km/l)	X 2.825	= Miles per gallon, Imperial (mpg)
Miles per gallon, US (mpg)	X	0.425	= Kilometres per litre (km/l)	X 2.352	= Miles per gallon, US (mpg)

Temperature

Degrees Fahrenheit = (°C x 1.8) + 32

Degrees Celsius (Degrees Centigrade; °C) = (°F - 32) x 0.56

*It is common practice to convert from miles per gallon (mpg) to litres/100 kilometres (l/100km), where mpg (Imperial) x l/100 km = 282 and mpg (US) x l/100 km = 235

Index

Printed and bound by CPI Group (UK) Ltd, Croydon, CR0 4YY

16/04/2025

14658480-0001